must be returned or renewe
ow      e loan period may be shorte
othe      der. A fine will be due if it is r

**DATE OF RET**

1992

17

# PRESSURE SENSORS

# MECHANICAL ENGINEERING

## A Series of Textbooks and Reference Books

**Editor: L.L. FAULKNER**   Columbus Division, Battelle Memorial Institute, and Department of Mechanical Engineering, The Ohio State University, Columbus, Ohio

**Associate Editor: S.B. MENKES**   Department of Mechanical Engineering, The City College of the City University of New York, New York

1. Spring Designer's Handbook, *by Harold Carlson*
2. Computer-Aided Graphics and Design, *by Daniel L. Ryan*
3. Lubrication Fundamentals, *by J. George Wills*
4. Solar Engineering for Domestic Buildings, *by William A. Himmelman*
5. Applied Engineering Mechanics: Statics and Dynamics, *by G. Boothroyd and C. Poli*
6. Centrifugal Pump Clinic, *by Igor J. Karassik*
7. Computer-Aided Kinetics for Machine Design, *by Daniel L. Ryan*
8. Plastics Products Design Handbook, Part A: Materials and Components; Part B: Processes and Design for Processes, *edited by Edward Miller*
9. Turbomachinery: Basic Theory and Applications, *by Earl Logan, Jr.*
10. Vibrations of Shells and Plates, *by Werner Soedel*
11. Flat and Corrugated Diaphragm Design Handbook, *by Mario Di Giovanni*
12. Practical Stress Analysis in Engineering Design, *by Alexander Blake*
13. An Introduction to the Design and Behavior of Bolted Joints, *by John H. Bickford*
14. Optimal Engineering Design: Principles and Applications, *by James N. Siddall*
15. Spring Manufacturing Handbook, *by Harold Carlson*
16. Industrial Noise Control: Fundamentals and Applications, *edited by Lewis H. Bell*
17. Gears and Their Vibration: A Basic Approach to Understanding Gear Noise, *by J. Derek Smith*

68.  Centrifugal Pump Clinic, Second Edition, Revised and Expanded, *by Igor J. Karassik*

69.  Practical Stress Analysis in Engineering Design, Second Edition, Revised and Expanded, *by Alexander Blake*

70.  An Introduction to the Design and Behavior of Bolted Joints, Second Edition, Revised and Expanded, *by John H. Bickford*

71.  High Vacuum Technology: A Practical Guide, *by Marsbed H. Hablanian*

72.  Pressure Sensors: Selection and Application, *by Duane Tandeske*

### Additional Volumes in Preparation

Zinc Handbook: Properties, Processing and Use in Design, *edited by Frank Porter*

### Mechanical Engineering Software

Spring Design with an IBM PC, *by Al Dietrich*

Mechanical Design Failure Analysis: With Failure Analysis System Software for the IBM PC, *by David G. Ullman*

# PRESSURE SENSORS

## SELECTION AND APPLICATION


**DUANE TANDESKE**

Sensym, Inc.
Sunnyvale, California



MARCEL DEKKER, INC.

New York • Basel • Hong Kong

Library of Congress Cataloging-in-Publication Data

Tandeske, Duane.
  Pressure sensors : selection and application / Duane Tandeske.
    p.    cm. --(Mechanical engineering ; 72)
  Includes bibliographical references and index.
  ISBN 0-8247-8365-4
  1. Pressure transducers. I. Title. II. Series: Mechanical
engineering (Marcel Dekker, Inc.) ; 72.
TJ223.T4T36  1990
681'.2--dc20                                              90-49147
                                                             CIP





Marcel Dekker, Inc.
270 Madison Avenue, New York, New York  10016

Current printing (last digit):

10  9  8  7  6  5  4  3  2  1

PRINTED IN THE UNITED STATES OF AMERICA

*To Sharon*

# Preface

<br><br>

As the use of microprocessors and computers increases, the need for sensors that can interface to computers also increases. Semiconductor technology has provided a way to produce low-cost, improved performance sensors that convert pressure inputs into proportional electrical output. This handbook discusses pressure-sensor fundamentals necessary to specify and apply sensors in practical applications. The technical information has been selected to discuss the sensors without getting too involved in semiconductor physics and sensor design techniques.

The first three chapters cover basic pressure measurement and how the development of microprocessors is increasing the demand for compatible sensors. Also, a discussion of the three types of pressure measurement should clear up some of the confusion as to what type of sensor should be used for each application. Chapter 4 gives a brief description of the most common transduction methods. The fabrication techniques used in piezoresistive sensors are covered in Chapter 5. Chapters 6, 7, and 8 cover specifications and selection of pressure sensors. Chapter 8 discusses the different techniques available to interface the sensors to microprocessors and computers, as well as discussing auto-zero and auto-referencing techniques using computers. Chapters 9 through 11 cover the more common applications of today's sensors. Finally, Chapter 12 looks at trends in pressure sensors.

In summary, this handbook provides the basic knowledge necessary to select and apply the pressure sensors available today.

*Duane Tandeske*

# Contents

# PRESSURE SENSORS

# 1

# A Need for New Sensors

## 1.1 TRADITIONAL PRESSURE MEASUREMENTS

Pressure-measuring devices can be divided into two groups: those in which the pressure is the only source of power, and those that require external electrical power. In the first group, the pressure that is being measured alters the shape or position of a mechanical element, which causes a pointer to be moved over a calibrated scale. A variety of mechanical elements have been used for this purpose, including (1) bellows, (2) diaphragms, and (3) Bourdon tubes.

### 1.1.1 Gages Using Bellows Elements

Figure 1.1 shows a cross section of a metal bellows attached to a pointer. As the pressure changes, the bellows expands, causing the pointer to move up the scale. As shown in Figure 1.1, the bellows is made from a thin-walled metal tube that has been formed with deep convolutions and sealed at one end. The stroke, or distance that the end of the bellows will move with a given pressure, is proportional to the number of convolutions and to the outside diameter. Bellows are frequently used if low pressures are to be measured. Bellows are made from elastic metal alloys such as brass, beryllium copper, and stainless steel; the application determines which material should be used. Chemical resistance and pressure range must be considered when designing a bellows element.

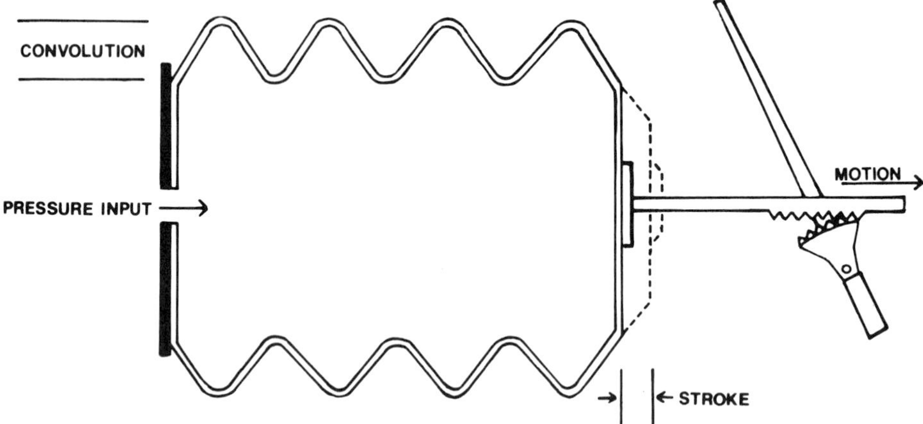


FIGURE 1.1   Bellows.


### 1.1.2   Gages Using Diaphragm Elements

The second sensing element is the diaphragm. Two basic configurations are used in pressure measurements: flat and corrugated.
Figure 1.2 shows a basic flat diaphragm; this configuration would
be used in a system that requires a minimum stroke. The corrugated
diaphragm contains a number of concentric corrugations, which
increase the effective area of the diaphragm and increase the stiff-
ness. The corrugated diaphragm allows a larger stroke than does
the flat diaphragm. Two corrugated diaphragms that have been
hermetically sealed together as shown in Figure 1.3 are frequently
referred to as an *aneroid* or *capsule*. To increase the effective
stroke, more than one capsule can be connected together, as shown
in Figure 1.4. This configuration of multiple capsules will have
the same characteristics as a much larger single capsule, which
makes it possible to construct low-pressure gages and still maintain
a reasonably small size.


### 1.1.3   Gages Using Bourdon Tube Elements

The third sensing element is the Bourdon tube, which is named
after the French inventor, Eugene Bourdon, who patented it in
1849. Bourdon elements are available in a number of configurations:
spiral, helical, twisted, and C-shaped. The Bourdon tube is made
from a section of oval tube that has been formed into a specific

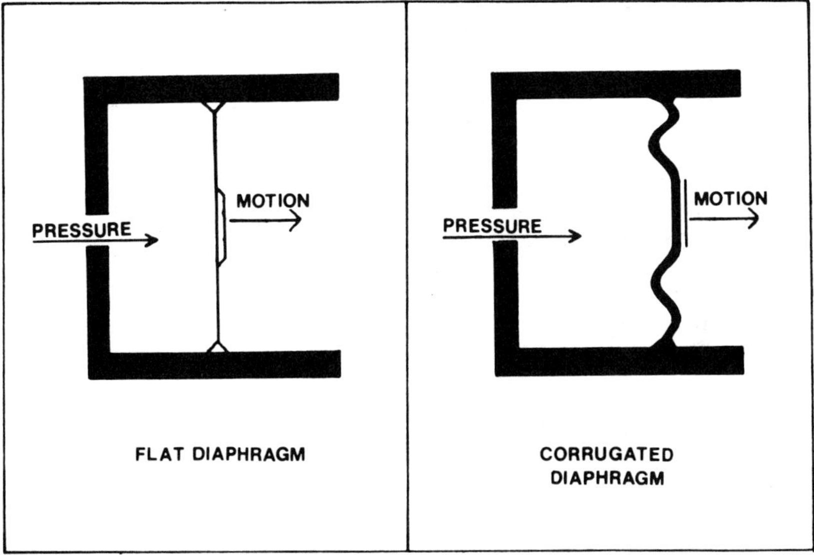


FIGURE 1.2  Diaphragm.

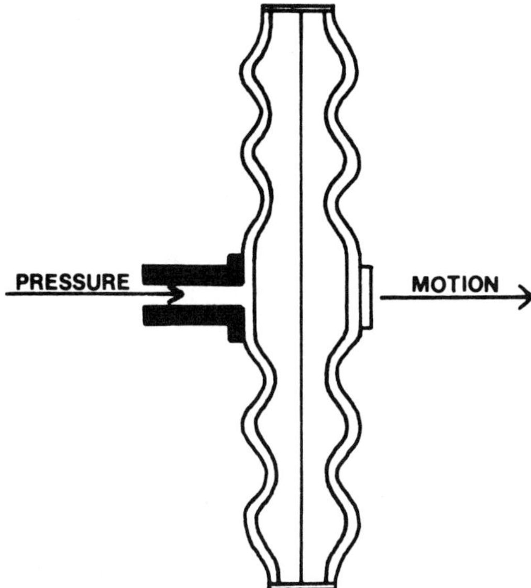


FIGURE 1.3  Capsule.

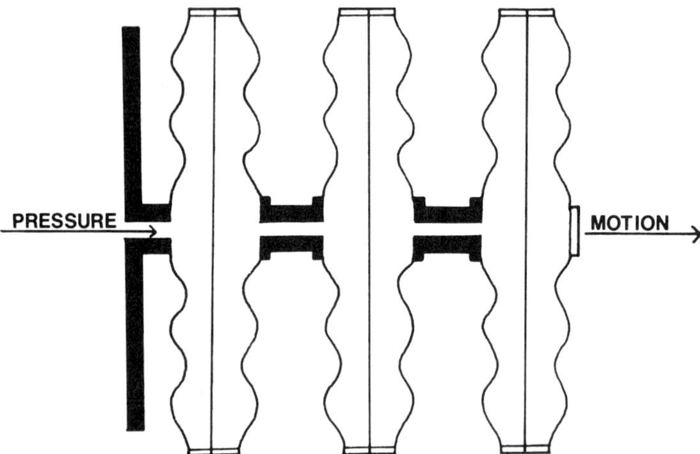


FIGURE 1.4   Aneroid.


shape. When pressure is applied to the tube, it tends to straighten.
This movement at the end of the tube is used to move a pointer
over a calibrated dial. The deflection of a Bourdon tube varies
with the ratio of its major-to-minor cross-sectional axes, the tube
length, the difference between the internal and external pressure,
and the total angle measured from its center of curvature. The
deflection is inversely proportional to the wall thickness of the
tube and the modulus of elasticity of the tube material after processing. Alloys such as brass, bronze, 300-series stainless steel, and
beryllium copper are commonly used to manufacture Bourdon tube
gages. The four types of Bourdon tubes are shown in Figure 1.5.
The C-shaped design is the most common. The spiral is similar
to the C-shaped tube in deflection characteristics; the tip travel
is proportional to the ratio of the total angle of curvature. This
type of gage is generally not used below 15 psi.


## 1.1.4   Pressure Manometer

Another popular pressure-measuring device is the manometer, which
dates back to the mid-1600s. The manometer is used as a measuring
device and as a standard to calibrate other measuring devices.
The basic manometer principle of hydrostatic balance is inherently
accurate—a column of liquid of known weight is balanced against
the unknown pressure. Reading the change in the column height
on a fixed scale indicates the amount of pressure applied. The

manometer is used to measure low pressures—from a column of water of less than an inch—and high pressures—up to 32 inches of mercury. Two fluids commonly used in manometers are mercury and oil. Oil, with a specific gravity of 0.8, is lighter than water and will move 1.2 times father than water in response to a given pressure. Mercury, which is 13.6 times the weight of water, will move 0.074 the distance in response to the same pressure. For this reason, oil is used for low-pressure manometers (inches of water), while mercury is used for higher pressures. Manometers are available in a variety of sizes and shapes, including U-tubes, cistern barometers, and slant tubes.

In its simplest form, the manometer is a U-tube, as shown in Figure 1.6. When a pressure is applied to one side of the U-tube, the liquid is forced into the other side of the tube until the weight

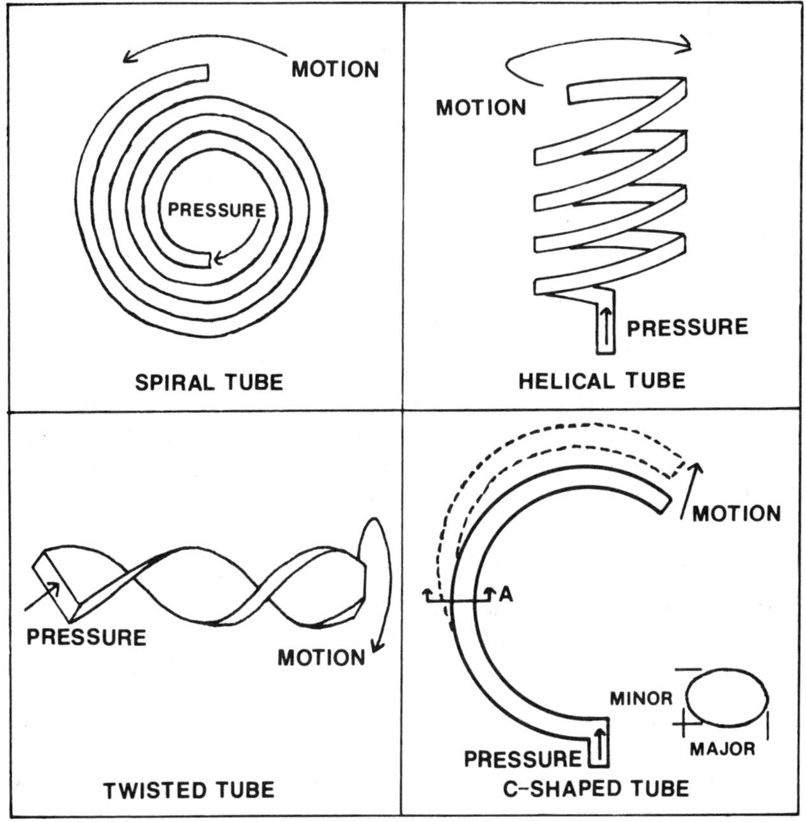


FIGURE 1.5  Bourdon-tube elements.

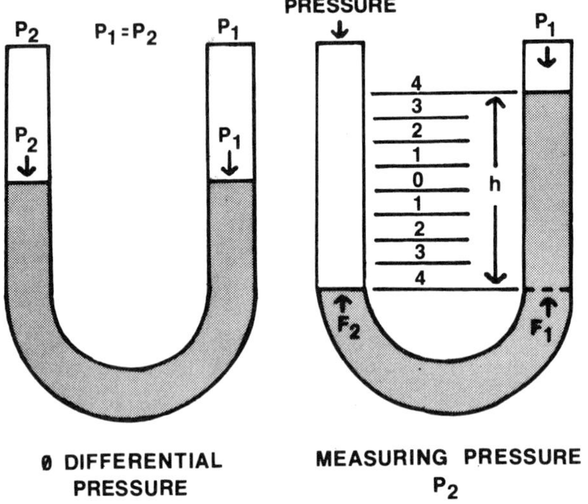



FIGURE 1.6   U-tube manometer.

of the column equals the pressure applied. The pressure is then
determined by measuring the difference in height of the two columns.
Pressure is defined as force per unit area, $P = F/A$, and every
physics student is familiar with the force equation, $F = MA$.

Consider a U-shaped tube with a uniform bore of area A and
half filled with a liquid of known mass. The manometer basic equation
is derived as follows:

At balance, $F_1 = F_2$. The upward force $F_1 = P_2 A$ and the
upward force $F_2 = pgh + P_1 A$, where p = density, g = gravity,
and h = column height. It follows that $P_2 A = pgh + P_1 A$.
Thus, the basic manometer equation is

$$P_2 - P_1 = pgh/A = pgh \text{ (per unit area A)}$$

Cistern barometers or manometers are designed so that the
diameter of one mercury column is larger than the other, forming
a reservoir that is usually called a cistern. A scale mounted along
the side of the tube is used to measure the height of the mercury
column. On barometers, this scale is made adjustable to compensate
for altitude. One standard atmosphere at sea level is 29.92 inches
of mercury; at 1000 feet above sea level, one atmosphere is 28.856
inches of mercury; and at 10,000 feet above sea level, one atmosphere
is 20.577 inches of mercury. To standardize, barometric pressure
is always normalized to sea level.

Another type of manometer is the slant tube or inclined design. To improve resolution on very-low-pressure manometers, the indicating tube is inclined to cause a greater linear movement than would a vertical tube (as shown in Figure 1.7). This type of manometer is used to measure pressures ranging from 0 to 2 inches of water to 0 to 10 inches of water.

### 1.1.5  Pressure Transducers

All the pressure gages discussed so far are self-contained and need no external power or equipment to function. Another type of sensor requires electrical power and a device to convert the electrical output to a usable form. As with mechanical gages, there are many different sensing elements used in pressure transducers, each one having certain advantages in specific applications. Pressure

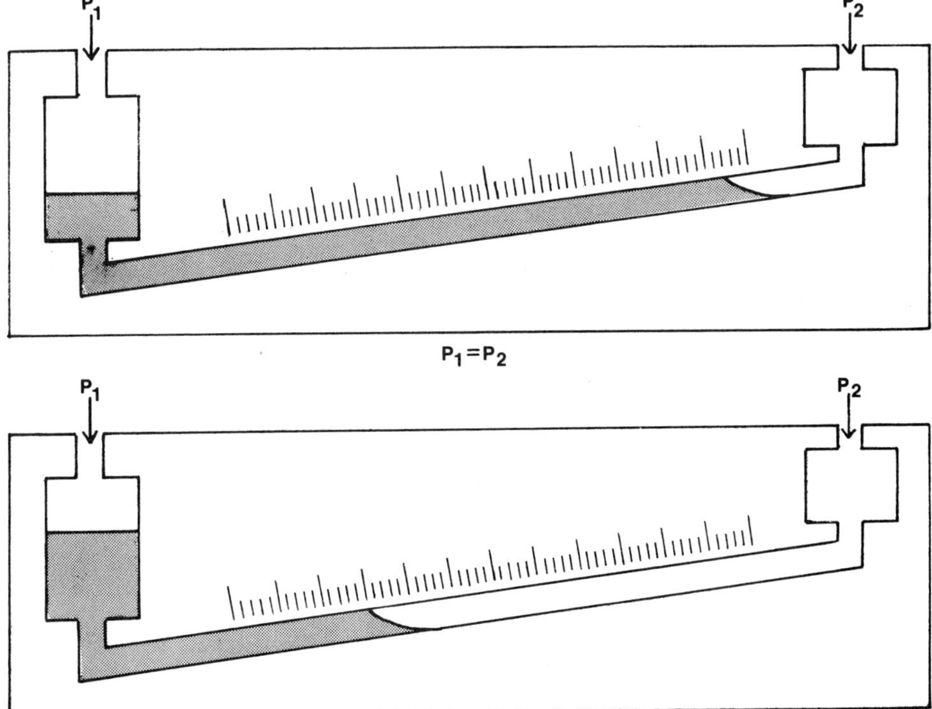


FIGURE 1.7   Inclined manometer.

range, accuracy, size, and price are a few of the parameters that determine which transducer to use for a specific application. Some of the more common transducer types are capacitance, force balance, potentiometric, bonded strain gage, variable reluctance, piezoelectric, differential transformer, and piezoresistive.

The outputs of most sensors available today are analog voltages. These voltages can be amplified and used to drive another piece of equipment, such as a strip-chart recorder or digital panel meter. The electrical output may be used to drive electronic recording equipment or a display at some distant location. These sensors also can be used in automatic control systems. For example, in some of today's cars, a sensor is used to measure the barometric pressure and automatically adjust the spark advance and carburetor as the car is driven at different altitudes. These transducers will be discussed in more detail in Chapter 4.


## 1.2  MICROPROCESSORS EXPAND MARKET

The microprocessor is a universal tool and can be designed into many different types of equipment, offering much more design flexibility than was available previously. A microprocessor can be programmed to carry out control functions as well as math and data storage. The basic microprocessor has three parts: the central processing unit (CPU), input/output (I/O) devices, and external memory. The heart of the microprocessor is the CPU, which contains the timing and control circuits, internal memory (called *resident memory*), and the arithmetic logic unit (ALU) (see Figure 1.8). These units, as well as any external units, are controlled by the timing and control unit in the CPU. Two interconnecting networks called *buses* are used to transfer instructions and data from one unit to another. The address bus accesses the unit to be addressed and transfers operating instructions from the control circuits. The data bus transfers the data from one unit to another.


### 1.2.1  Types of Memory

One source of confusion when talking about microprocessors is that the word *memory* can mean either read-only memory (ROM) or random-access memory (RAM). The ROM memory is permanent or nonvolatile, and the microprocessor cannot change the contents of the memory, just read it. The RAM memory is used by the microprocessor, and data can be written into memory as well as read. Every time the power is turned off, all RAM memory is cleared and all data are

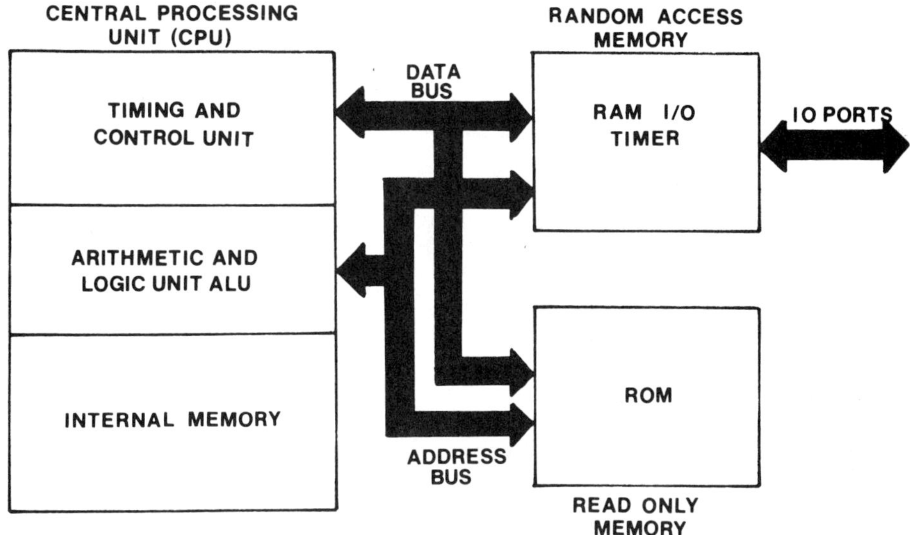


FIGURE 1.8   Basic microprocessor.

lost; the ROM memory, being nonvolatile, remains unchanged. Both
RAM and ROM are in the microprocessor or computer hardware
and can be expanded by adding more RAM or ROM chips. Another
type of memory is floppy disks or tape drives external to the microprocessor or computer. These devices can be used to store large
amounts of data. The computer controls the operation of the tape
system and can read or write onto the disk.

## 1.2.2   Basic Microprocessor System

In the simplest system, the operating program is stored in memory
(ROM) and, when power is turned on, the microprocessor automatically
reads the program and carries out the operating sequence. The
operating program is a set of instructions that the microprocessor
reads from ROM and carries out line by line. A simple example
is a sprinkling system that will perform the following functions:
(1) reset system, (2) read soil moisture, (3) check system, and
(4) turn sprinkler on and off.
   The sequence of events that the microprocessor will perform
is called the *flowchart*. From the flowchart, a flow diagram is drawn
to show the logic steps necessary to perform each function. Figure 1.9
shows the flow diagram for this system. Next, a program is written:

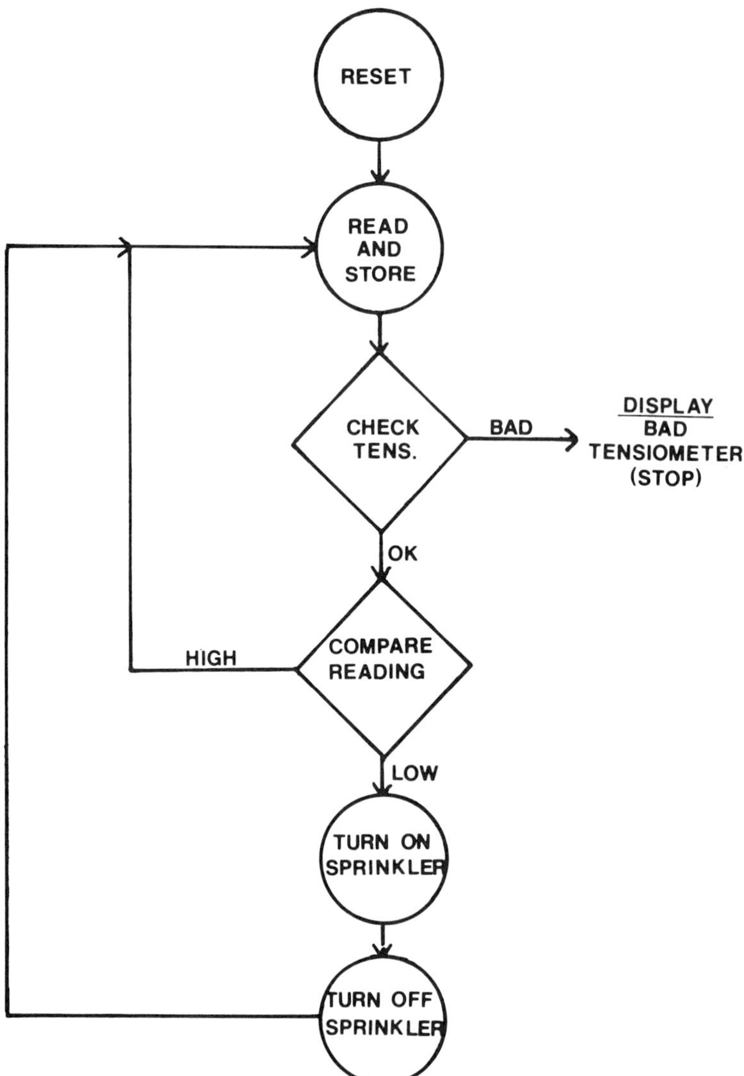



FIGURE 1.9  Sprinkler system flow diagram.

1. Reset
2. Read tensiometer (soil moisture)
3. Store reading in A
4. If A > 20 go to 11
5. If A < 5 go to 11
6. If A > 10 go to 2
7. Turn on sprinkler
8. Wait 10 minutes
9. Turn off sprinkler
10. Go to step 2
11. Output (check tensiometer)
12. Stop

When the system is turned on, it will reset and then read the output of a tensiometer, a device that measures the amount of moisture in the soil and outputs a linear voltage proportional to the moisture content. The reading is stored in RAM and is called A (line 3). Each time a reading is taken, A will be updated. Lines 4 and 5 check the tensiometer output to ensure that it is in a reasonable range and the tensiometer is working properly. If the tensiometer appears to be malfunctioning, the program jumps to line 11 and outputs a warning signal to check the tensiometer. If the reading is in range, line 6 compares the reading to the limit (10). If the reading is greater than 10 (water content is high), the program jumps to line 2 and starts over again. If the reading is less than 10 (water content is low) the program jumps to line 7 and turns on the sprinkler. Line 8 and 9 let the sprinkler run for 10 minutes and then turn it off. Line 10 starts the sequence all over again by jumping to line 2. When the system is turned on, it will continue to run until it gets a check-tensiometer signal and stops at line 12 or is turned off. This is a simple control system, and one microprocessor could control many such systems if the program were expanded.

## 1.2.3 Expanding Markets

In the past 10 years, the use of the microprocessor has expanded to include such markets as automobiles and appliances. When microprocessors are used in a control application, some sort of sensing element is used to provide feedback. In applications like appliances and automobiles, the microprocessor needs a variety of sensors to perform its control functions. Microprocessors are being used to control the engine timing on cars and trucks, by measuring such things as engine temperature, revolutions per minute (RPM), altitude, and engine vacuum. On cars that use a carburetor, three pressure sensors are used to measure altitude, manifold vacuum, and air density. On cars that use a fuel-injection system, an altitude sensor will be required. The automotive market is very large (millions of units per year).

Microprocessors and sensors are showing up in other markets as well, such as appliances like washing machines, refrigerators, and water heaters. In washing machines, functions like measuring water level, dispensing soap, and weighing clothes can be performed by a microprocessor and three sensors. An automatic washing machine would have a transducer built into the lid that weighs the clothes and feeds these data to the microprocessor, which then determines the appropriate amount of soap and water that should be used. Another sensor measures the water level as the tub is filled, and a third dispenses the appropriate amount of soap. Both these applications require pressure transducers that have electrical outputs that can be used by the microprocessor.

By creating new applications, the microprocessor is increasing the size of the electronic sensor market. The traditional mechanical sensor is still used in applications that require a visual pressure indication. One impact that the microprocessor is having on the sensor market is the price of sensors. Microprocessor chips can be purchased for 5 to 10 dollars in small quantities. (If the price of the sensors were 60 or 70 dollars, the design project would probably be canceled.) Today, temperature-compensated and calibrated sensors are just starting to get into the price range that makes it practical to design a system using a microprocessor and multiple sensors. This makes a lot of designs practical that used to be too expensive.


## 1.3  COMPUTERS NEED FEEDBACK

Computers, like microprocessors, require data input through a keyboard or from external devices such as temperature or pressure sensors to perform control functions. Computers are used in large office buildings to decrease power consumption by monitoring heating and air-conditioning systems. When temperature sensors are placed throughout the building, the computer can determine which rooms need cooling or heating and can open dampers, allowing only those rooms outside the desired temperature range to be affected. The computer is also used to monitor the condition of the filters in these systems by measuring the pressure drop across them. At the right time, the computer displays a message to change the filters or automatically start a cleaning process. This improves efficiency by ensuring a minimum impedance to the airflow. Another industry that is using computers to improve efficiency is the process-control industry.

### 1.3.1 Process Control

The process-control industry is being influenced by computer technology. New plants being constructed are using computers to automatically control the manufacturing process. Traditionally, all the sensor outputs are transmitted to a central control room, where all the variables are displayed on gages arranged on a control panel. An operator monitors the panel and makes necessary adjustments or walks through the plant monitoring the various gages and making necessary adjustments in the field. In the computer-based systems, most of the gages have been replaced by cathode-ray tube (CRT) screens. All the process variables are programmed into the computer, which then scans the various sensors measuring the process and automatically adjusts the various temperatures and pressures throughout the process. Having a continuous control system can improve the quality of the product by decreasing variations in the process. The computer can keep detailed records of process variables and provide the data necessary to optimize facilities management and future expansion.

### 1.3.2 Process-Control Transmitters

The process-control industry has developed its own type of transducers, called *transmitters*, which are designed for the special needs of the industry. Because most processing plants are large and spread out, signals have to be transmitted long distances to the control room. The industry has adopted a current signal as the standard for transmission: 4 milliamperes (mA) represents zero and 20 mA represents full scale. The current loop is not affected by wire impedance and noise, as are voltage output devices. One-process-control problem is that large numbers of transmitters a long distance from the control room require miles of wire. To solve this problem, the two-wire transmitter was developed. This transmitter reduces the number of wires per transmitter from three to two. The two-wire transmitter uses the same two wires to power the device and transmit the signal (see Figure 1.10). A voltage—usually 24 or 28 volts (V)—is applied to the sensor, and the loop current is measured across a low-impedance load. Some transmitters offer optional features such as special housings for hazardous environments or operate over a range of pressures. Being able

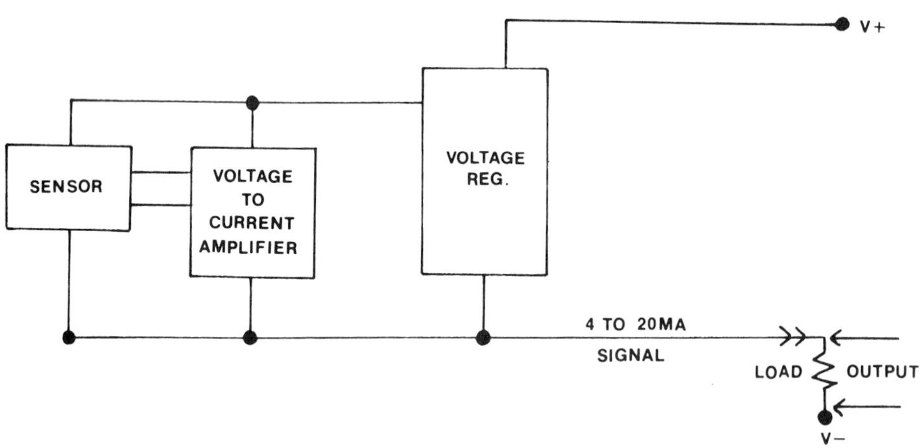


FIGURE 1.10  Process control transmitter.

to adjust the sensitivity of the transmitter is called *turndown*. For example, a transmitter that operates from 2 psi full scale to 20 psi full scale is said to have a 10-to-1 turndown. The variable-gain feature allows one transmitter to be set to any full-scale pressure between 2 and 20 psi, thus reducing the number of transmitter pressure ranges required.

## 1.3.4  Industrial Applications

When you walk into a hardware store or a large insurance company, there seems to be a computer terminal at every desk and counter. The computer terminal is as popular as the typewriter in most offices. Most industrial computer applications are word processors and data-handling systems. In hardware stores, the computer is used for inventory control—it keeps track of the quantity of parts on hand and when a replacement order should be placed. In insurance companies, the computer is used to keep track of clients and do statistical calculations.

These computers range from desktop computers to large mainframe systems. The large mainframe systems can be time-shared to do many functions at once, giving many terminals access to the computer at the same time. These same systems used in factories are called controllers and have all the same basic functions. The major difference is that the controller has to interface with the physical world; to do this, special I/O ports are added. For example, when a conveyor belt moves containers through a dispensing machine, the computer needs sensors to determine the position of the container

and its weight. The computer also will have to be able to turn
the dispenser on and off, as well as control the speed of the conveyor belt. The computer needs two input signals and two outputs
to perform this simple control function. A variety of input signals
can be used, depending on the type of function the computer is
to perform:

| Function | Device |
| --- | --- |
| Count objects | Switch input: light beam or mechanical switch |
| Measure speed | Linear input: count a frequency from a tachometer or read the output of a hall-effect sensor mounted on a gear or pulley |
| Read pressure | Pressure transducer and analog-to-digital converter |
| Temperature | Temperature sensor and analog-to-digital converter |
| Weight/force | Load cell and analog-to-digital converter |

The switch and frequency inputs are connected to the I/O ports,
and the computer is programmed to scan these ports to determine
if the input is high or low. The analog-to-digital converters are
connected to the data bus through a holding register or can be
read serially through an I/O port. The computer also needs to
control functions by outputting commands to various hardware:

| Function | Device |
| --- | --- |
| Start/stop | Special I/O port that drives electronic relay to turn on and off AC (alternating current) power |
| Control flow | Switch output to turn valves on and off or linear output through a digital-to-analog converter |
| Speed control | Linear output to motor control logic |

The I/O ports output logic signals to valves and electronic
relays. Digital-to-analog converters are controlled from holding
registers that are updated by the data bus.


## 1.3.5  Home Computers

In the home, computers are used as word processors, as calculators,
and to play games. The possible applications around the home are

limited only by the imagination of the user and the availability of the necessary hardware and software. Some of the possible applications include control functions such as heating only occupied rooms, watering only when the soil in the yard has reached a predetermined moisture level, and controlling water heaters and solar heaters for maximum efficiency. All these functions are possible if I/O boards are available to allow the computer to interface with other devices. These boards are available from companies that have specialized in computer interface products, and are advertised in the commercial trade magazines. Unfortunately, these devices are expensive and not easy to use; they require some modifications and additions to the basic system. As the computer industry matures, more and more interface products will be made available to the home-computer market.

## 1.4  SMART SENSORS

In the past few years, a new term has been introduced into the sensor market: *smart sensor*. A smart sensor is microprocessor based, meaning that it uses a microprocessor in the signal-conditioning electronics. Having the microprocessor available gives the sensor much more flexibility and allows the system designer to use a very versatile tool to design control loops.

### 1.4.1  Smart Transmitters

The leading transmitter manufacturers have introduced a series of devices called *smart transmitters* that are microprocessor based. The microprocessor can store data digitally to correct for sensor nonlinearities, temperature errors, and static pressure effects. Each sensor is tested over the working temperature and pressure range and these data are stored digitally in the microprocessor. As the sensor output is conditioned by the microprocessor, the microprocessor automatically compensates for static errors. Some features of these transmitters are high turndown (50-to-1 or higher), digitally adjustable damping from 0 to 32 seconds, temperature output options, and external digital programmability.

Smart transmitters have two output modes: One is the standard 4- to 20-mA current loop where the signal is converted from analog to digital and then to 4 to 20 mA, and the second is to transmit the digital signal. The first option can be used in the existing 4- to 20-mA systems. An interface card can be added to the system that imposes a digital signal on the same two wires, allowing the user to communicate with the transmitter through the host computer.

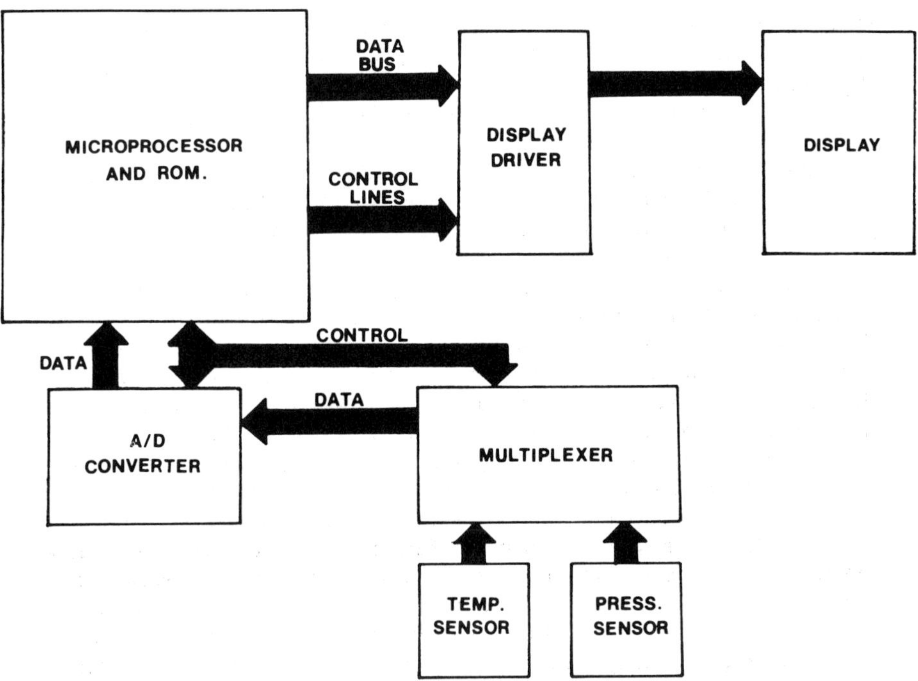



FIGURE 1.11   Basic system.

For example, the user could change the damping factor or pressure range of the sensor automatically at predetermined times in the process. The second option (digital output) will be the obvious choice if a computer is used to control the system. The analog-to-digital converter is no longer necessary to interface to the computer, eliminating one source of error.

These transmitters can be programmed by the host computer, as well as transmitting back current status for a system check. Manufacturers of smart transmitters offer accessories, such as a hand-held communicator that can be connected to the transmitter on site or from the control room to configure the transmitter parameters if a computer is not used. Interface cards are also available that contain all the hardware necessary to connect multiple smart transmitters to the host computer.

### 1.4.2   Smart Sensors in Altimeters and Barometers

Some applications require very specialized smart sensors. Altimeters and barometers require very accurate sensors to measure very small changes in pressure. To buy a sensor tested to the required accuracy would be cost prohibitive—the sensor would cost more than the finished barometer. For these applications, sensors need to have good repeatability but not necessarily tight specifications; temperature and linearity can be 0.5 to 1 percent. The sensors are tested over temperature and the resulting data are stored in a programmable ROM (PROM) serialized to match the sensor. A pressure sensor, temperature sensor, and microprocessor make up the finished unit.

In the case of the barometer, a keypad is made available to program the device for the desired pressure units and initial calibration data. The microprocessor can be programmed to store barometric pressure over a period of time and to display the changes over this period of time, as well as the current barometric pressure. A number of features can be programmed into the microprocessor.

The altimeter has a unique problem in that pressure vs. altitude is not a linear function. The linearity of the sensor is measured over the pressure range of interest, this error is algebraically added to the pressure-vs.-altitude function, and the resulting data are burned into a PROM. The basic system will look the same for both altimeter and barometer; each includes a pressure sensor, temperature sensor, and PROM containing correction and calibration data (see Figure 1.11). To compensate the sensor, the temperature data must be known. Each system that uses this technique can display temperature as well as barometric pressure with no additional hardware. The technique of sensor compensation gives the user a reasonably priced solution to sensor accuracy that was not available before the use of microprocessors.

# 2
# Pressure Fundamentals



## 2.1 PRESSURE MEASUREMENT OR LOAD CELL

Two common measurements that cause confusion are force and pressure. Unfortunately, in the United States, they both use the same units. It is possible to have a force of one pound as well as a mass of one pound and, if that isn't confusing enough, the unit of measure for pressure is pounds per square inch (psi). The technical community of the world has agreed upon a set of units called SI (System International) units, in which the basic units for mass, force, and pressure are as follows:

Unit of mass = *kilogram* (A cylinder of platinum-iridium alloy is kept at the International Bureau of Weights and Measures at Paris; a duplicate at the U.S. National Bureau of Standards serves as the mass standard for the United States.)
Unit of force = *newton* (The force that accelerates one kilogram by one meter per second squared.)
Unit of pressure = *newton per square meter* (The term pascal—abbreviated Pa—has been designated.)

The pascal is a very small pressure (0.000145 psi); most applications use the term *kilopascal* (abbreviated kPa) (1000 pascals), which is 0.145 psi.

### 2.1.1  What Is a Load Cell?

The difference between a load cell and a pressure transducer is
the following:

force = MA (mass × acceleration)

pressure = F/A (force/area)

Force-measuring instruments fall into two classifications:

1. Counterbalance
   a. Spring balance (bathroom scale)
   b. Mass balance (counterbalance scale)
2. Deflection (Stress) Force
   a. Strain-gage load cell
   b. LVDT (linear variable differential transformer)
   c. Hydraulic load cell


### 2.1.2  Load-Cell Applications

Force is applied to a load cell through a metal adapter (rod, eyebolt,
threaded coupling) or by attaching a platform to the load cell to
weigh objects. In each case, the force is applied to the body of
the load cell, resulting in a deflection or strain of the cell. The
hydraulic load cell is a pressure transducer with a special interface
that converts the applied force into a pressure that the sensor
can measure. By designing the equivalent of a piston connected
to a force-coupling device, a known pressure-to-force ratio is
established:

$$F_2 = F_1 \times A_1 / A_2$$

where $F_1$ is the force applied to piston of area $A_1$, $A_2$ is the area
of the connecting piston, and $F_2$ is the resulting force. For example,
if a 1-square-inch piston is connected to a 10-square-inch piston,
a ratio of 10 to 1 is established. A force of 100 pounds generates
a pressure of 10 psi at the sensor.

Some common applications of load cells include weight scales,
thrust measurement, and force measurement.


### 2.1.3  Pressure-Sensor Applications

A pressure sensor is used to measure pressure in a liquid or gas
such as air, water, or oil. The pressure is coupled to the sensor

through a tube, pipe, or the atmosphere. The sensor converts
the applied pressure into an electronic signal or a deflection that
moves a pointer. The pressure sensor will not respond to a force
applied directly to the sensor case.

Some common applications of pressure sensors include barometers,
water-pressure gages, air-pressure gages, oil-pressure gages,
and vacuum gages.

## 2.1.4  Types of Load Cells

The most common application of the load cell is the scale. Less
expensive scales (e.g., bathroom scales) use the force-balance
load cell (Figure 2.1). More expensive scales (e.g., doctor's scales)
use the counterbalance load cell (Figure 2.2). In the force-balance
load cell, a platform is mounted on a network of springs that are
calibrated to allow a certain deflection for an applied force. The
deflection is coupled to a pointer or moving scale that indicates
the applied force. The deflection could also be coupled to a potentiometer or LVDT, which converts the deflection to an electronic
signal that is displayed electrically. Two types of strain-gage load
cells are bending beam and shear beam. To maximize output signal,
strain gages are bonded to the body of the load cell in a Wheatstone
bridge configuration. A force is applied to the load cell, resulting
in a deflection of the beam that creates a strain in the bonded
strain gages.

## 2.1.5  Strain Gages

The term *bonded strain gage* can apply to nonmetallic (semiconductor)
gages or metallic (wire or foil) gages. The metallic gage is the
most common; it has been used for years and is well characterized.
Comparing the gage to a single straight length of conductor (small-
diameter wire), if the wire is strained in the longitudinal axis, the
resistance increases proportionally to the strain. As the wire is
stretched, the length increases and the diameter decreases, increasing the impedance. To increase the effective length of the resistor,
gages are designed in a grid pattern and are available in a variety
of sizes and shapes. Each strain gage is given a number to represent
its strain sensitivity (gage factor). When strain gages are manufactured, a sample from each lot is measured for gage factor and the
average is marked on the package. The gage factor (GF) is

$$GF = \Delta R/R/\Delta L/L$$

where R = resistance and L = length.

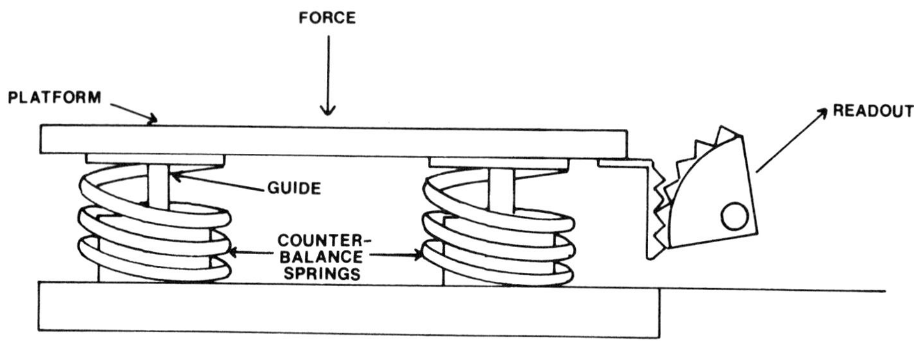


FIGURE 2.1  Force-balance load cell.

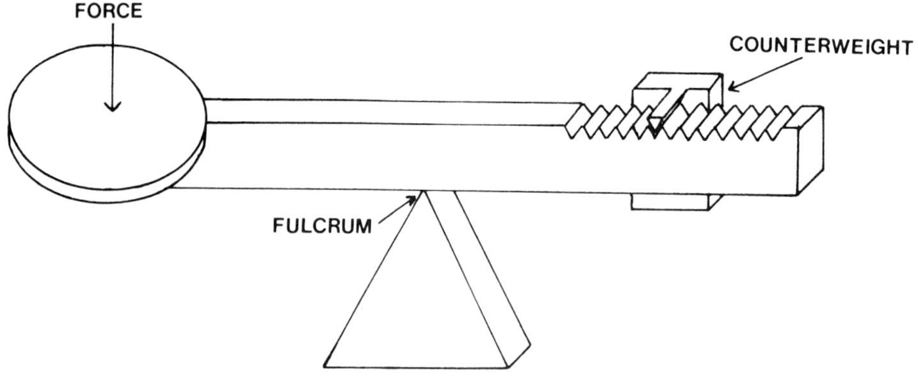


FIGURE 2.2  Counterbalance scale.

Gage factor is a dimensionless quantity; the larger the value, the more sensitive the strain gage. Figure 2.3 shows the pattern of the basic foil strain gage. The gage is designed to be sensitive to strain in the longitudinal axis and has minimum sensitivity to strain transverse to this axis. Sensor sensitivity is increased by placing one or more strain gages in a Wheatstone bridge configuration. When the sensor is strained, the Wheatstone bridge is unbalanced, producing a differential output voltage. To maximize sensitivity, two resistors are positioned to be in compression and two others to be in tension (Figure 2.4). Bonded-strain-gage sensors have an output of 50 millivolts (mV) to 150 mV full scale. The basic metallic strain gage is a grid-pattern resistor bonded to a thin insulating material. Multiple resistors can be bonded to one piece of insulating

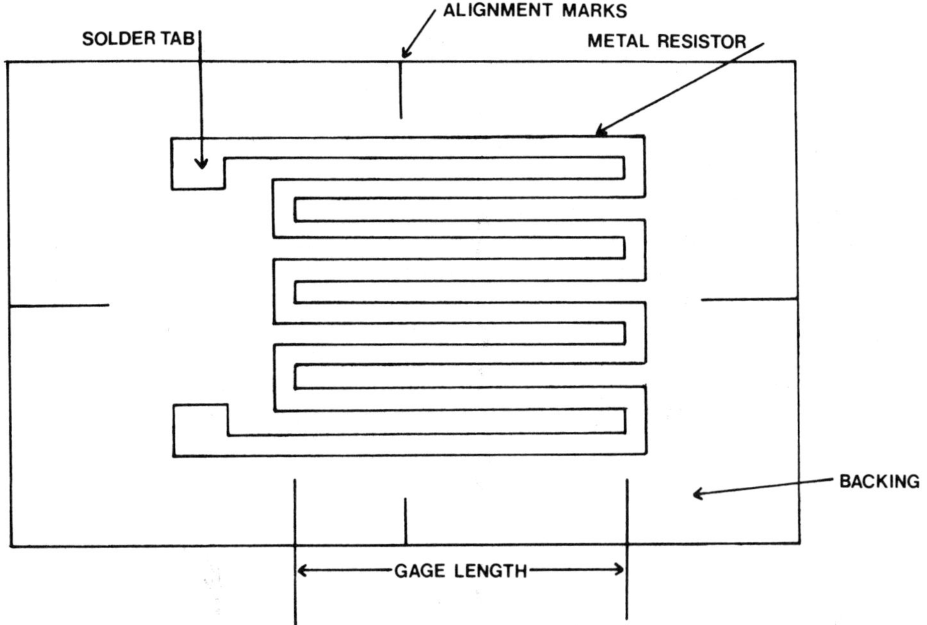


FIGURE 2.3   Foil strain gage.

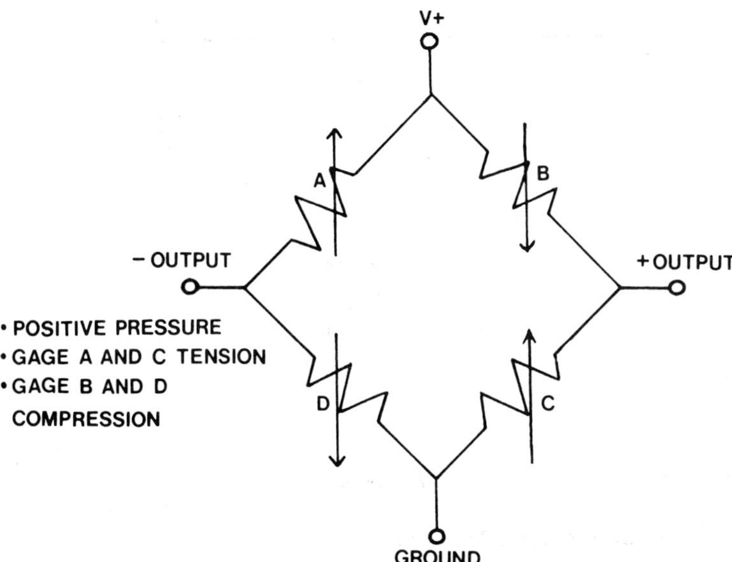


FIGURE 2.4   Wheatstone bridge.

material in a variety of patterns to measure strain in more than
one axis. Bonded metallic strain-gage sensors typically have 120 to
360-ohm impedance and gage factors of approximately 2.

## 2.1.6  Bending-Beam Load Cell

The bending-beam load cell is very popular due to its simple design
and low cost. Basically, the load cell is a rectangular metal beam
with strain-gage sensors bonded to the top and bottom of the beam
to measure applied force. The beam is secured at one end and the
force is applied to the other end. The strain gages measure the
strain at the center of the beam. The gages bonded to the top
of the beam are in tension, while the gages bonded to the bottom
are in compression. Figure 2.5 shows the basic bending-beam load
cell.

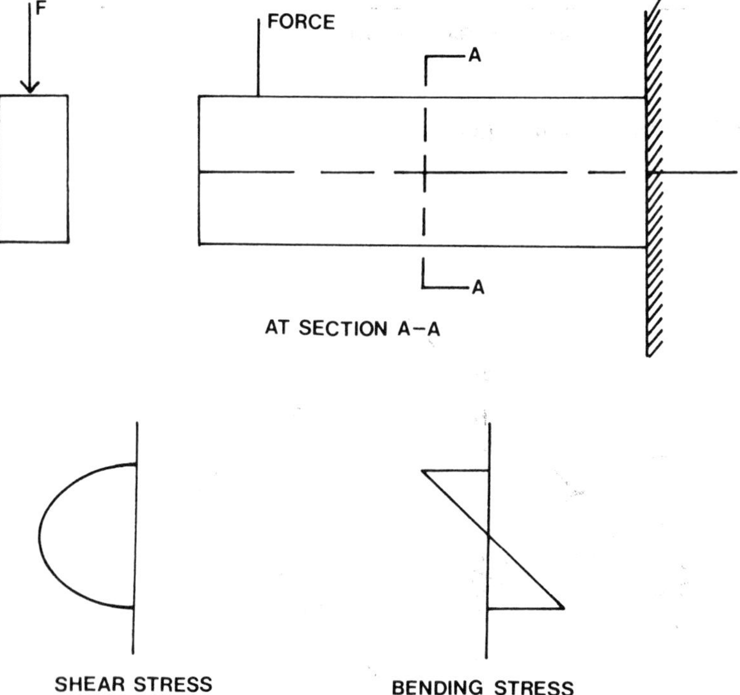


FIGURE 2.5  Bending-beam load cell.

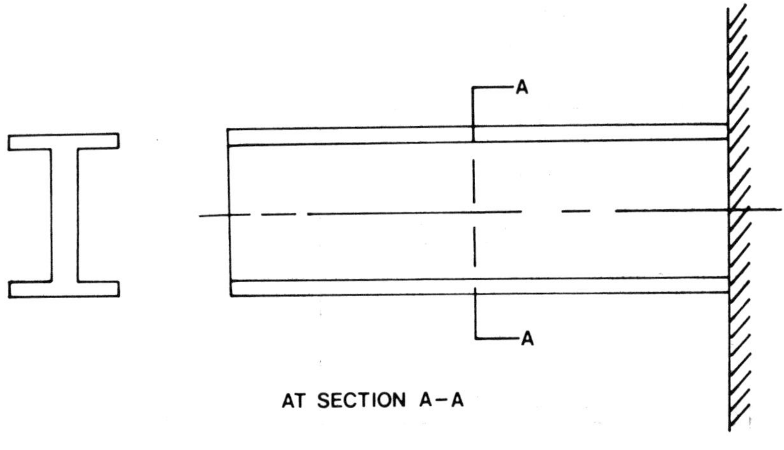

AT SECTION A-A

SHEAR STRESS                    BENDING STRESS

FIGURE 2.6   Shear-beam load cell.

## 2.1.7   Shear-Beam Load Cell

The shear-beam load cell has a number of distinct advantages, including high side-load rejection, faster return to zero after load, and higher tolerance of dynamic forces and vibration. The shear-beam principle measures the shear forces produced by an applied load. Figure 2.6 demonstrates that the shear stresses on a bending beam are constantly changing. However, the I beam produces a shear stress that can be measured by strain gages mounted on the web of the beam. Shear-beam load cells are available in a variety of sizes and shapes. One popular configuration is the S shape shown in Figure 2.7. This shape is easy to use, and the load is easy to mount.

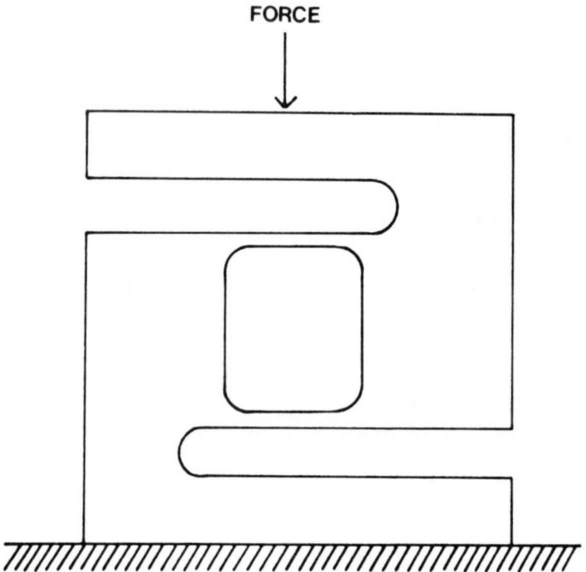


FIGURE 2.7   Shear-beam load cell construction.


## 2.2   PNEUMATICS

When you feel the wind in your face or smell a fragrance from across
the room, you are experiencing matter in the gaseous state. We
experience the effects of a gas without seeing it. The earth is
surrounded by the atmosphere, a layer of air contained by the
earth's gravitational forces. Since gravitational forces decrease
with distance, the atmosphere is denser near the earth's surface
than at higher altitudes. These pressure variations are used to
measure altitude or to predict weather patterns. A quantity of gas
can be stored indefinitely in a closed container; if the container
is opened, the gas is heard escaping, but it diffuses into the
atmosphere and cannot be seen. To better understand pneumatic
pressure measurements, the properties of gases will be discussed.


### 2.2.1   Properties of Gases

Gas consists of molecules in random and rapid motion constantly
colliding with each other or anything in their vicinity. The gas
molecules (or atoms) are so small and so far apart that the volume
occupied by the particles is negligible to the total volume. There

are no attractive forces between the molecules, and they repel each other upon collision. The motion of the molecules provides them with kinetic energy. ($KE = 1/2\ mv^2$, where m is mass and v is velocity.) The collisions result in no loss of kinetic energy, but kinetic energy can be transferred from one molecule to another. The rapid motion and lack of attractive forces of the gas molecules explain why a gas occupies the total volume of the container in which it is placed. The molecules continue to move in an outward direction until they collide with the inside of the container. The number of collisions per unit area determines the pressure the gas exerts on the container. The gas is easily compressed into a smaller volume due to the large distance between the molecules.

## 2.2.2 Boyle's Law

Boyle's law states that the volume of a sample of a gas at a constant temperature is inversely proportional to the pressure. This can be expressed algebraically as

$$V = k/P$$

where k is a proportionality constant, or

$$PV = k$$

The product of P and V is a constant for a sample of gas at a constant temperature.

If the pressure is changed, the product of the new pressure and volume will equal the same constant:

$$P_1 V_1 = k$$
$$P_1 V_1 = PV$$

Solving for the final volume,

$$V_1 = V(P/P_1)$$

*Example*: A piston is compressing 2 liters of gas in a cylinder at 100 kPa. If the pressure is increased to 140 kPa, what is the new volume?

$$V_1 = 2(100/140)$$
$$V_1 = 1.4286\ \text{liters}$$

When the pressure is increased 29 percent, the volume decreases by 29 percent.

The same equation can be solved for a change in pressure due to a change in volume:

$$P_1 = P(V/V_1)$$

*Example*: Using the same numbers as in the previous example, 100 kPa compressing 2 liters of gas, when the volume is decreased to 1.4286 liters, what is the resulting pressure?

$$P_1 = 100(2/1.4286)$$

$$P_1 = 140 \text{ kPa}$$

2.2.3  Charles's Law

Charles's law states that the volume of a sample of a gas at constant pressure is directly proportional to the Kelvin temperature:

$$V = kT$$

$$k = V/T$$

If the temperature is changed,

$$V_1/T_1 = k$$

$$V/T = V_1/T_1$$

Solving for the final volume gives

$$V_1 = V(T_1/T)$$

All the gas laws use Kelvin temperature to avoid 0 or negative numbers. Kelvin is degrees Celsius plus 273. For example, 25°C is 298 K (25 + 273).

*Example*: What volume is occupied by a 2.5-liter sample of gas at a constant pressure when the temperature is increased from 25°C to 50°C? The temperatures must be converted to Kelvin, 25°C = 298 K and 50°C = 323 K. Then,

$$V_1 = 2.5(323/298)$$

$$V_1 = 2.710$$

The volume increases from 2.5 to 2.710 when the temperature is increased 25°C.

### 2.2.4  Pressure-Temperature Law

The pressure-temperature law states that the pressure of a sample of gas of fixed volume is directly proportional to the Kelvin temperature:

$$P_1 = P(T_1/T)$$

*Example*: A tank of gas with a pressure gage has a pressure of 2 bars at a temperature of 25°C. What would the pressure be at −10°C?

$$P_1 = 2(263/293)$$

$$P_1 = 1.795 \text{ bars}$$

A decrease in temperature of 35°C causes a pressure change of −0.205 bar. The average kinetic energy of the gas molecules is proportional to the temperature. Kinetic energy is $1/2 \, mv^2$; as the temperature is decreased, the kinetic energy and velocity decrease or, if the temperature is increased, the kinetic energy and velocity will increase proportionally.

### 2.2.5  Combined Gas Law

The combined gas law is a combination of Boyle's and Charles's laws. The combined gas law states that the change in volume is inversely proportional to pressure and directly proportional to Kelvin temperature:

$$V_1 = V(P/P_1)(T_1/T)$$

*Example*: If a balloon is filled with 2.5 liters of gas at 1 atmosphere (atm) and 25°C, what is the volume if the pressure is increased to 1.5 atm and the temperature is increased 50°C?

$$V_1 = 2.5(1/1.5)(348/298)$$

$$V_1 = 2.5(.6667)(1.1678)$$

$$V_1 = 1.9464 \text{ liters}$$

The volume decreases because the increase due to the rise in temperature is less than the decrease due to the increase in pressure.


### 2.2.6  The Perfect Gas Law

The perfect gas law states

$$PV = nRT$$

where P is pressure, V is volume, n is the number of moles of gas (*mole* is defined as the amount of a substance that contains Avogadro number, $6.023 \times 10^{23}$, molecules), R is the universal gas constant, and T is temperature Kelvin.

$$R = (0.0821 \text{ liter atm/K mole})$$

R has the same value for all ideal or perfect gases. The constant relates the product of the temperature and number of moles of gas to the product of the pressure and volume.

The perfect gas law is quite useful: If three of the properties associated with the gas sample are known, the fourth can be calculated easily. The volume of the sample is usually expressed in liters. The pressure is usually expressed in atmospheres, and temperature is always expressed in Kelvin degrees. The perfect gas law (or the ideal gas law) can be rearranged to solve for any one of the four terms:

$$V = nRT/P \text{ for volume}$$

$$P = nRT/V \text{ for pressure}$$

$$T = PV/nR \text{ for temperature}$$

$$n = PV/RT \text{ for number of moles}$$

*Example*: A weather balloon contains 5 moles of helium. What will be the volume at an altitude of 30,000 ft., a temperature of -45°C (228 K), and a pressure of 0.2974 atm?

$$V = nRT/P$$

$$V = (5 \text{ moles}/0.2974 \text{ atm})(0.0821 \text{ liter atm/K mole})(228 \text{ K})$$

$$V = 314.7 \text{ liters}$$

Note that moles, atm, and K all cancel and the final volume is in liters.

### 2.2.7  Properties of Gases Summary

The properties of volume, pressure, and temperature are dependent
on one another and can be related.

*Volume*:  A gas is in constant random motion, filling any container
   in which it is placed. Gases are easily compressed due to the
   large distances between molecules. Volume is inversely proportional to pressure. Volume is proportional to temperature (Kelvin).
*Pressure*: The pressure a gas exerts on the container depends on
   the number of collisions per unit area, which in turn depends
   on the number of molecules and how fast they are moving.
   Pressure is proportional to the temperature of the gas (kinetic
   energy or speed of the molecules). Pressure is inversely proportional to the volume.
*Temperature*: Temperature is always expressed in degrees Kelvin
   ($^\circ$C + 273). Temperature determines the kinetic energy of the
   gas molecules and the speed at which they are moving.

### 2.2.8  Vacuum Measurements

A perfect vacuum is defined as the absence of all gas molecules.
Any pressure less than atmospheric is considered a partial vacuum.
A vacuum can be measured relative to atmospheric pressure (vacuum
gage) or relative to a hard vacuum (absolute). Hard vacuums are
measured in torr (mmHg) or millitorr. A vacuum is generated by
pumping gas out of a closed container. Space is a natural vacuum.

## 2.3  HYDROSTATICS

Like gases, liquids are in constant motion, but the molecules are
closer together and intermolecular attraction holds the molecules
in a more condensed state.

### 2.3.1  Liquid State

The liquid state consists of a collection of particles in a condensed
state. The intermolecular forces are greater in liquids than in gases
due to the close proximity of the molecules. Unlike gases, liquids
are considered noncompressible. At most pressures, they will assume
the shape of the container, but will not expand to fill the entire
volume of the container. Some gases, such as nitrogen, can be
liquefied and stored in a liquid state.

### 2.3.2  Properties of Liquids

As a gas is cooled, the average kinetic energy of the molecules decreases. At high pressures, the molecules move closer together and the attractive forces become important. Under proper conditions of temperature and pressure, a gas can be liquefied. Liquefaction occurs when the attractive force between molecules overcomes kinetic motion, so that the molecules accumulate into the liquid state. Liquid air is at a temperature of -190°C. The liquid state is dynamic, and the molecules are moving about randomly with short distances between collisions. Because of the freedom of movement and the attractive forces that exist between molecules, all liquids can flow and will assume the shape of the container in which they are placed. However, the attractive forces do exhibit a certain internal resistance to flow, which is called *viscosity*. Some liquids, such as oils, can be quite viscous, while others have very low viscosities and flow readily. The viscosity of a liquid depends on the nature of its particles. The viscosity can be increased by cooling, since the attractive forces rather than kinetic forces are dominant. Like gases, the average kinetic energy is proportional to the Kelvin temperature of the liquid.

### 2.3.3  Water

Pure water is a colorless, odorless, tasteless liquid that boils at 100°C and freezes at 0°C. Water is a vital component of living systems. It is the only chemical that is found in all three states naturally—ice, water, and water vapor. Ninety-seven percent of the earth's water is in the oceans, while only 1 percent is fresh water in the soil and lakes. Water is constantly evaporating to form clouds and falling to earth as rain or snow. One of the most remarkable properties of water is its ability to act as an excellent solvent. Many compounds are readily dissolved in water

### 2.3.4  Pascal's Law

Pascal's law states that a fluid in a container transmits pressure equally in all directions. Blaise Pascal, a French scientist and mathematician, developed the law during the 1600s. If a container is filled with a liquid and a 1-psi pressure is applied to the top of the liquid, the pressure of the water will increase by 1 psi over each square inch of the inside of the container.

### 2.3.5  Surface Tension

The attractive forces have a great influence on the behavior of
the liquid state. Within a liquid, a given molecule is attracted equally
by the molecules around it. The molecules on the surface of the
liquid are attracted only by the molecules beside and below them.
The unbalanced forces cause a net inward pull that tends to draw
the surface molecules into the body of the liquid. The surface of
the liquid is under strain or tension; this is called the *surface
tension* of the liquid. A needle or razor blade can be suspended
by the surface tension of water. If the needle or razor blade is
pushed below the surface of the water, it will sink.

### 2.3.6  Evaporation

All molecules do not have the same kinetic energy; kinetic energy
is transferred from molecule to molecule during collisions. At a
given temperature, the majority of the molecules have the same
kinetic energy, and a smaller percentage will have very high or
very low energy levels. The kinetic energy of the molecules tends
to oppose the intermolecular attractive forces. Some molecules that
possess sufficient kinetic energy and are near the surface can break
away from the attractive forces and escape to the vapor state. This
process is called *evaporation*. After a rain shower, small pools of
water evaporate in a matter of minutes. Increasing the temperature
of the liquid increases the average kinetic energy, and more molecules
will have sufficient energy to escape. Since the molecules with higher
energy are the ones that evaporate, the liquid's average kinetic
energy and temperature decrease. Water evaporating from the skin
makes you feel cooler. The amount of evaporation depends on the
amount of surface area of the liquid: A shallow puddle of water
will evaporate completely, while a bottle of water with a small opening
stays in the liquid state for long periods of time.

### 2.3.7  Vapor Pressure

If water is placed in a closed container, the molecules that escape
into the vapor state by evaporation cannot escape from the container.
Under these conditions, a certain amount of liquid evaporates into
the vapor state; eventually, however, the proportion of vapor to
liquid in the container becomes constant. As more and more molecules
enter the space above the liquid, the pressure increases and the

number of collisions with the inside of the container and the surface
of the water increases, resulting in some of the molecules returning
to the liquid state. This change from the vapor state to the liquid
state is called *condensation*. The pressure of the vapor in dynamic
equilibrium with a liquid at a given temperature is called the *vapor
pressure* of the liquid (water-vapor pressure at 25°C is 23.7 torr).

## 2.3.8 Boiling

The vapor pressure of a liquid increases with an increase in temperature. If a quantity of water is heated in an open container, a point
is reached at which further heating does not increase the temperature
of the liquid, and bubbles of water vapor are formed rapidly throughout the entire liquid volume. These bubbles rise to the surface and
burst as the vapor escapes. Boiling will not take place until the
vapor pressure of the liquid is equal to the pressure of the surrounding atmosphere. Addition of more heat to boiling water will not
increase the temperature of the liquid, but will increase the rate
of evaporation or boiling. At sea level, water boils when heated
to a temperature of 100°C or 212°F. Boiling temperature is a function
of ambient pressure:

| Absolute pressure | Boiling point |
| --- | --- |
| 1 psia | 101.74°F |
| 14.696 psia | 212.0°F |
| 100.0 psia | 327.82°F |
| 1000.0 psia | 544.58°F |

## 2.3.9 Steam

Steam is created by heating water to the boiling point. The steam
is as hot as the boiling water. It requires 100 calories of heat to
raise one gram of water from the freezing point (0°C or 32°F) to
the boiling point. To change the same gram of boiling water into
steam takes an additional 540 calories of heat. This heat is called
the steam's *latent heat*. It is released when the steam changes back
to liquid water. Steam fills more space than the water from which
it comes. In a closed vessel, large increases in pressure occur
as the water is turned into steam. At this stage, it is called *saturated
steam*. If heated more, it becomes superheated steam and generates
even higher pressure. Steam is used in thermodynamic systems for
heat transfer and power generation.

## 2.4  HYDRODYNAMICS

We have considered gases and liquids at rest and briefly discussed
some of the laws governing the properties of each. Hydrodynamics—
the study of liquids in motion—will be covered next. (The term
*liquids* can mean gases or liquids.) There are three basic laws
of hydrodynamics: the principle of continuity, Bernoulli's law, and
Torricelli's theorem. These laws apply to steadily flowing liquids.

### 2.4.1  Principle of Continuity

The principle of continuity in fluid flow states that the velocity
of a liquid flowing through a pipe increases as the inside diameter
of the pipe decreases and decreases as the inside diameter increases.
The nozzle on a garden hose uses this principle: As the area of
the nozzle is reduced, the velocity of the water increases.

### 2.4.2  Bernoulli's Law

Bernoulli's law states that the pressure of a fluid increases as its
velocity decreases and decreases as its velocity increases. Bernoulli's
law is used in the design of flow meters, as well as in the design
of an airplane wing. This law applies to both gases and liquids.
As air passes over the curved surface of a wing, the velocity increases. This produces less pressure on the top than on the under-
side of the wing, creating lift (see Figure 2.8).

### 2.4.3  Torricelli's Theorem

Torricelli's theorem states that the velocity with which a fluid flows
through an opening in a container equals the velocity of a body
falling from the surface of the liquid to the opening. Torricelli's
theorem does not apply to gases, because a gas has no surface.

### 2.4.4  Laminar Flow

Before discussing the principles and equations of flow-measurement
techniques, it is appropriate to define the two possible types of
flow: laminar (viscous or straight line) and turbulent (eddy type).
In laminar flow, the path of each particle of fluid is a straight line
parallel to the walls of the pipe. The turbulent flow path consists

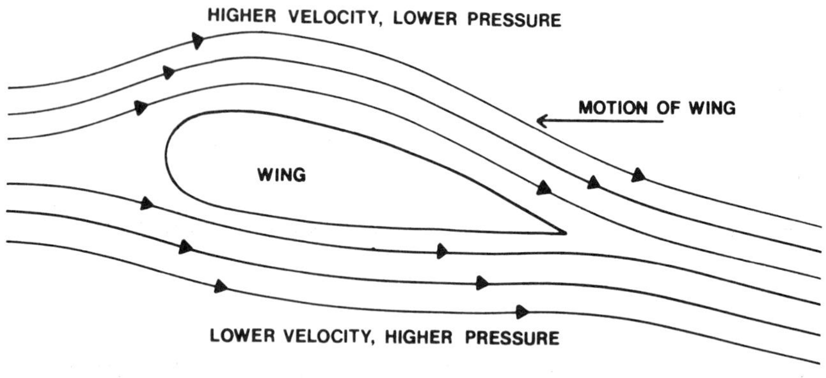


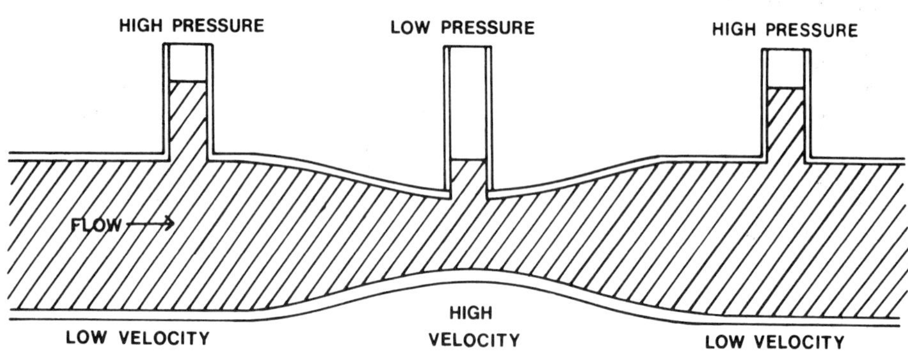


FIGURE 2.8   Bernoulli's law.

of eddies and curls. These two types of flow are shown in Figure 2.9.
The energy relations and, hence, the measurement principles are
different for laminar and turbulent flow. A turbulance factor, known
as the Reynold's number, can be derived for each flow application.

## 2.4.5   Reynold's Number

Whether the flow is laminar or turbulent depends on four factors:

1.   Velocity of the fluid V (the higher the velocity, the more turbu-
     lent the flow)

2. Density of the fluid, $\rho$ (the higher the density, the more turbulent the flow)
3. Diameter of the passage, D (the larger the diameter, the more turbulent the flow)
4. Viscosity of the fluid, $\mu$ (the greater the viscosity, the less turbulent the flow)

Using these four factors, a turbulence factor can be defined as

$$R_n = VD\rho/\mu$$

This factor is known as the Reynold's number—and it is, surprisingly, dimensionless. The numerical value of this important factor is the same in English units, or metric units, or any consistent set of

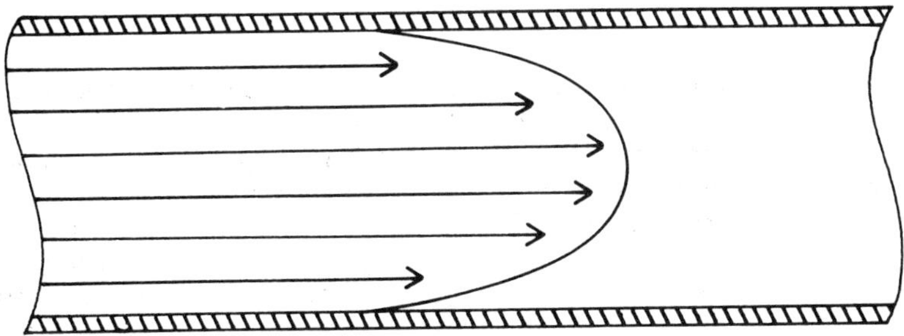

**LAMINAR FLOW**
**(REYNOLD'S NUMBER < 2000)**

**TURBULENT FLOW**
**(REYNOLD'S NUMBER > 4000)**

FIGURE 2.9  Types of flow.

units. A Reynold's number above 4000 indicates a turbulent flow;
below 2000, the flow is always laminar. In the 2000-to-4000 range,
the flow is neither purely laminar nor turbulent, and the user must
apply all formulas with caution. Although the Reynold's number
itself is dimensionless, each factor has its own units. In the SI
system,

$V$ = meters per second (m/s)

$D$ = meters (m)

$\rho$ = kilograms per square meter (kg/m$^2$)

$\mu$ = kilograms per meter second (kg/m · s)

Note that the acceleration at gravity (standard) $g$ = 9.80 m/s$^2$.


### 2.4.6  Types of Flow Meters

Numerous types of flow meters are available for closed piping systems.
We will limit this discussion to the differential pressure devices,
which include orifices, venturi tubes, flow nozzles, and pitot tubes.
Differential flow meters are the most common type in use today.
The basic operating principle of differential pressure flow meters
is based on the premise that the pressure drop across the meter
is proportional to the square of the flow rate. The flow is determined
by measuring the pressure differential and taking the square root.
A differential flow meter contains a device that causes a change in
kinetic energy, which creates the differential pressure, and a device
to measure the pressure and create a readout or display that represents the flow value.


### 2.4.7  Orifices

The orifice is simply a washer-shaped piece of metal with a precisely
sized hole in the center. This unit is clamped into a section of
pipe between a set of flanges, as shown in Figure 2.10. The orifice
is low cost, ±2 percent to ±4 percent accurate, and has a medium
pressure loss. Pressure taps on either side of the orifice measure
the differential pressure variations across the device.


### 2.4.8  Venturi Tubes

Venturi tubes have the advantage of having less pressure drop
at high flow rates. As shown in Figure 2.11, the venturi tube is

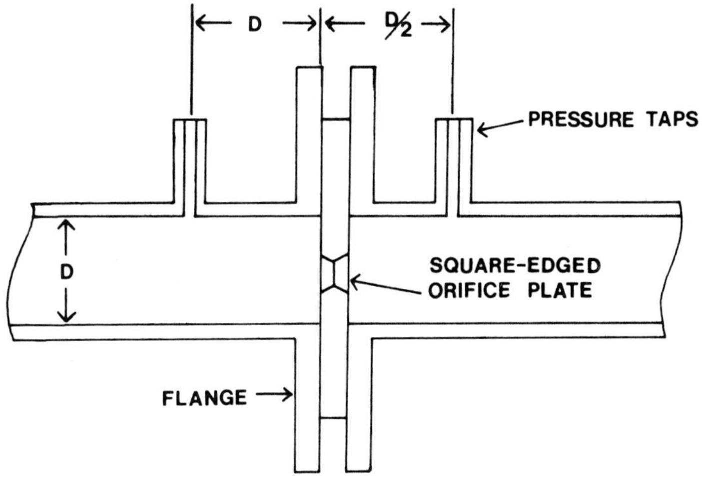



FIGURE 2.10   Orifice.

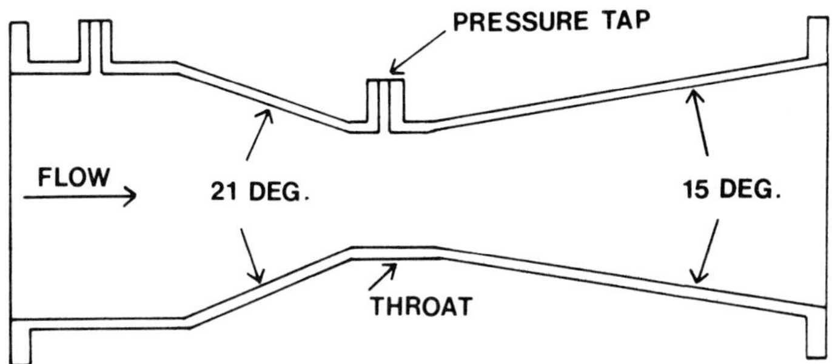



FIGURE 2.11   Venturi.

a section of pipe that has a tapered entrance, a straight throat, and a tapered exit section. Pressure taps are installed in the entrance section and the throat. As the fluid is accelerated through the narrower throat section, a pressure differential is developed between the input and output regions. The carburetor on any car is a venturi-based device.

## 2.4.9   Flow Nozzles

Flow nozzles are similar in operation to orifice plates. The flow nozzle is used to measure higher flow rates that have the same

pressure drop. As shown in Figure 2.12, the flow nozzle has a
smooth entrance section and a small-diameter section to generate
the pressure drop.


## 2.4.10   Pitot Tubes

Pitot tubes measure two pressures simultaneously: impact and static
pressure. The impact pressure is measured by a tube bent at a
right angle and positioned in the direction of the flow. The static
tube is positioned to be at a right angle to the direction of flow.
The pitot tube consists of an impact tube that receives the flow-
related pressure inside a larger tube that receives the static pressure
from radial sensing holes around the tip (see Figure 2.13). Figure 2.14
shows the three common types of combined pitot tubes in use today.
The two pressures developed at the pitot tube are measured with
a differential pressure transducer. The transducer automatically
compensates for any change in static pressure. For example, referring
to Figure 2.13, if a no-flow condition exists, $P_1$ equals $P_2$ and the
sensor output is zero. If the duct is blocked, restricting the airflow,
$P_1$ and $P_2$ both increase, but the sensor output is still zero. If
the duct is unblocked, $P_1$ will increase due to the change in flow
and $P_2$ will measure the new static pressure. Pitot tubes are well

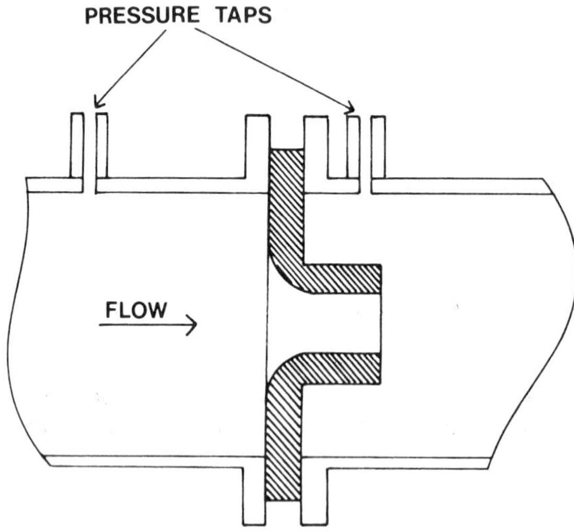


FIGURE 2.12   Flow nozzle.

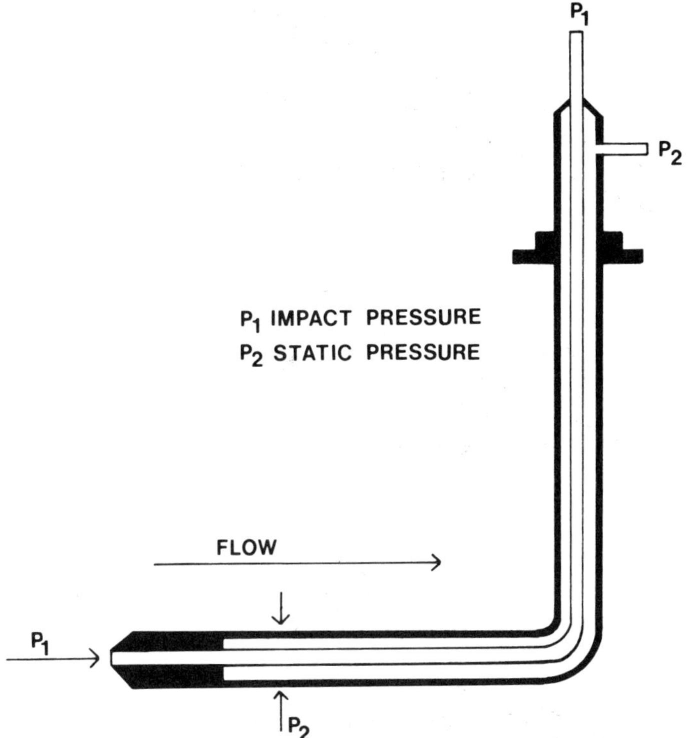


FIGURE 2.13   Pitot tube.

adapted to a variety of applications, such as measuring the airspeed of an airplane or the flow through a heating or ventilation duct.

## 2.4.11   Calculating Air Velocity

When a pitot tube is used to measure velocity in a closed pipe, the flow near the wall is retarded by friction and moves at a slower rate than does the air at the center of the stream. If the pitot tube is placed near the wall of the pipe, a minimum reading is measured by the pitot tube. As the pitot tube is moved across the diameter of the pipe, the readings increase and peak at the center of the pipe. A traverse can be made to determine the average flow velocity. A traverse is made by taking a number of readings across the diameter of the pipe and using the average. The center can be determined by making a traverse and multiplying the result

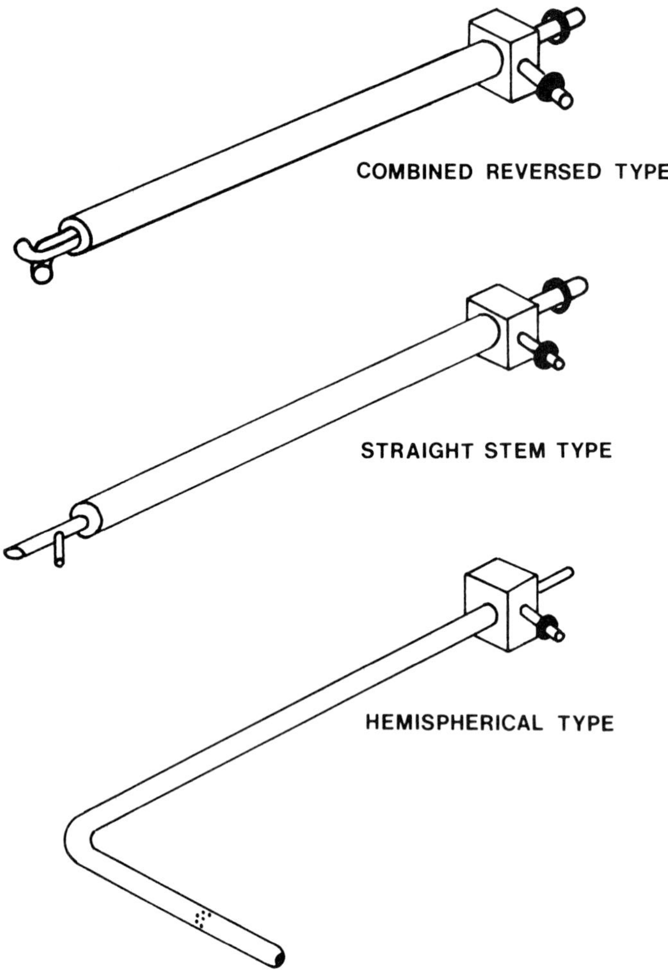


FIGURE 2.14   Types of pitot tubes.

by 0.89 for gas or air to determine the average velocity. Positioning
the pitot tube is very critical in measuring airflow: The dynamic
pressure port must be positioned directly into the airstream, and
the flow should be laminar (free of elbows, size changes, or obstruc-
tions upstream from the pitot tube).

The velocity for hemispherical and straight-stem pitot tubes
is calculated using the basic formula

$$V = 1096.5 \, hv/d \text{ (for gases and liquids)}$$

where  $V$ = velocity in feet per minute

$hv$ = velocity pressure in inches of water (pitot-tube reading

$d$ = density of the air in pounds per cubic foot

Dry air density is calculated by

$$d = 1.325 \, P_6/T$$

where  $d$ = air density in pounds per cubic foot

$P_6$ = barometric pressure in inches of mercury

$T$ = absolute temperature (°F plus 460)

Combined reverse-type pitot tubes use the formula

$$V = 976.98 \, hv/d \text{ (airflow and liquids)}$$

## 2.4.12  Volume Calculations

To measure volume flow, the average air velocity and cross-sectional area of the pipe or duct must be known:

$$Q = AV$$

where  $Q$ = quantity of flow in cubic feet per minute

$A$ = cross-sectional area of the duct in square feet

$V$ = average velocity in feet per minute

## 2.5  PRESSURE UNITS

Pressure is defined by two types of units: force per unit area, or height of a liquid column that can be supported by a given pressure (also called *head pressure*).

## 2.5.1  Force per Unit Area

Any combination of force divided by a unit of area results in a unit of pressure. Combinations such as pounds per square inch, pounds per square foot, or pounds per square yard are possible.

The more common units of force per unit area are pounds per square inch (psi) or newtons per square meter (pascal). One kilopascal (1000 pascal) equals 0.145 psi.

    pressure = force/area

From the equation, the larger the area to which a force is applied, the less the pressure generated from a given force.

    force = pressure × area

Conversely, a very small pressure applied over a large area will develop large forces. One pound per square inch is 144 pounds per square foot or 1296 pounds per square yard. Houses are destroyed in a tornado due to sudden changes in air pressure on the outside of the house. This change in pressure is relatively small; however, the outside area of the house is large, generating high forces from inside the house. Units of force per unit area are used when operating in moderate to high pressure ranges (10 to 10,000 psi).


## 2.5.2  The SI System

The SI system (Système International) has been adoped by 36 countries, including the United States. The basic units for mass, force, and pressure are

Mass = kilogram (a cylinder of platinum–iridium alloy kept at the
    International Bureau of Weights and Measures in Paris)
Force = newton (the force that accelerates one kilogram by one
    meter per second squared)
Pressure = newton per meter squared (pascal)

The unit for pressure is the pascal (0.000145 psi) or the kilopascal (1000 pascals). Another unit used in the SI system is the bar; one bar is 100 kilopascals (14.5 psi). The bar is useful for high pressures; for example 1000 psi = 6,896,551 pascals, 6896 kilopascals, or 68.96 bars. The SI system is less confusing when dealing with force and pressure measurements at the same time: kilogram = mass, newton = force, and kilopascal = pressure.


## 2.5.3  Liquid Head Pressure

The second type of pressure unit is height of liquid, or head pressure. This unit is different from force per unit area in that the

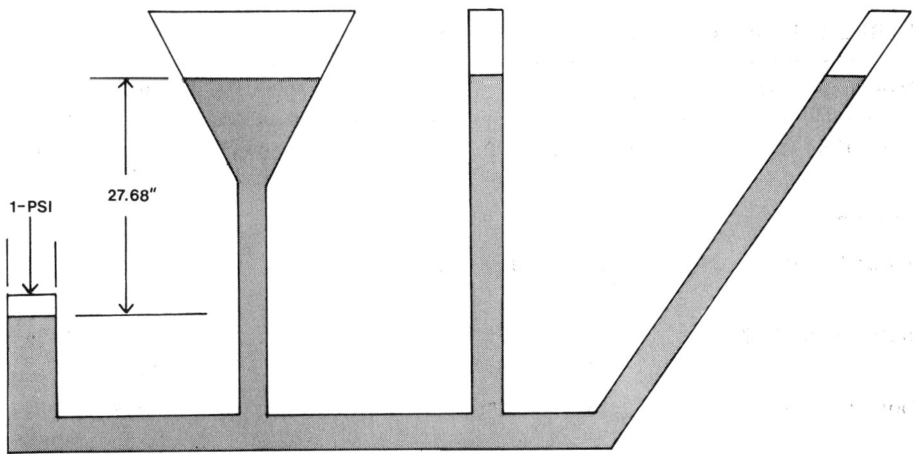


FIGURE 2.15   Static head pressure.

pressure is independent of applied area. Small pressures are measured in liquid head-pressure units, inches of water, inches of
mercury, and centimeters of water.

The height of liquid that can be supported by a given pressure
is determined by

$$P = D \times H$$

where P is the applied pressure

   D is density of the liquid

   H is height of the liquid

Density of the liquid is temperature dependent and is given
at a reference temperature. The head-pressure equation contains
no area term, since the area or shape of the tube has no effect
on the height of liquid a pressure will support. A pressure of
1 pound per square inch will support a column of water 27.68 inches
at a temperature of 39°F (see Figure 2.15). Very-low-pressure
sensors are calibrated in inches of water, some as low as 1 inch
full scale. Water and mercury are the two common liquids used
to measure pressure. Inches of mercury are used to measure atmospheric pressure in barometers, while inches of water are used in
flow and very low-pressure measurements.

TABLE 2.1  Pressure Conversion Chart

| Pressure unit | atm | psi | in $H_2O$ | ft $H_2O$ | in Hg | Pa |
|---|---|---|---|---|---|---|
| Std. atm | 1.0 | 14.696 | $4.0679 \times 10^2$ | 33.90 | 29.921 | $1.0133 \times 10^5$ |
| *English* | | | | | | |
| pound/sq in | $6.8045 \times 10^{-2}$ | 1.0 | 27.680 | 2.3067 | 2.0360 | $6.8946 \times 10^3$ |
| inch water 4°C | $2.4582 \times 10^{-3}$ | $3.6127 \times 10^{-2}$ | 1.0 | $8.3333 \times 10^{-2}$ | $7.3554 \times 10^{-2}$ | $2.4908 \times 10^2$ |
| foot water 4°C | $2.9499 \times 10^{-2}$ | 0.4335 | 12.00 | 1.0 | 0.8826 | $2.989 \times 10^3$ |
| inch mercury 0°C | $3.342 \times 10^{-2}$ | 0.4912 | 13.596 | 1.1330 | 1.0 | $3.386 \times 10^3$ |
| *SI Units* | | | | | | |
| pascal ($N/m$ ) | $9.869 \times 10^{-6}$ | $1.4504 \times 10^{-4}$ | $4.0147 \times 10^{-3}$ | $3.346 \times 10^{-4}$ | $2.953 \times 10^{-4}$ | 1.0 |
| kilopascal | $9.869 \times 10^{-3}$ | 0.145 | 4.0147 | 0.3346 | 0.2953 | $1.0 \times 10^3$ |
| bar | 0.986 | 14.504 | $4.0147 \times 10^2$ | 33.46 | 29.529 | $1.0 \times 10^5$ |
| millibar | $9.869 \times 10^{-4}$ | $1.4504 \times 10^{-2}$ | 0.4015 | $3.346 \times 10^{-2}$ | $2.9529 \times 10^{-2}$ | $1.0 \times 10^2$ |
| *Metric* | | | | | | |
| gram/sq cm | $9.6784 \times 10^{-4}$ | $1.4224 \times 10^{-2}$ | 0.3937 | $3.2809 \times 10^{-2}$ | $2.8959 \times 10^{-2}$ | 98.097 |
| kilogram/sq cm | 0.9678 | 14.224 | $3.9371 \times 10^2$ | 32.809 | 28.959 | $9.8097 \times 10^4$ |
| cm water 4°C | $9.6781 \times 10^{-4}$ | $1.4223 \times 10^{-2}$ | 0.3937 | $3.2808 \times 10^{-2}$ | $2.8958 \times 10^{-2}$ | 98.0626 |
| meter water 4°C | $9.6781 \times 10^{-2}$ | 1.4223 | 39.370 | 3.2808 | 2.8958 | $9.8063 \times 10^3$ |
| mm mercury 0°C | $1.3153 \times 10^{-3}$ | $1.9337 \times 10^{-2}$ | 0.5353 | $4.4605 \times 10^{-2}$ | $3.9370 \times 10^{-2}$ | $1.333 \times 10^2$ |

| kpa | bar | mbar | g/cm$^2$ | kg/cm$^2$ | cm H$_2$O | m H$_2$O | mm Hg |
|---|---|---|---|---|---|---|---|
| $1.0133 \times 10^2$ | 1.0133 | $1.0133 \times 10^3$ | $1.0332 \times 10^3$ | 1.0332 | $1.0333 \times 10^3$ | 10.333 | $7.6000 \times 10^2$ |
| 6.8946 | $6.895 \times 10^{-2}$ | 68.946 | 70.306 | $7.0306 \times 10^{-2}$ | 70.308 | 0.7031 | 51.715 |
| 0.2490 | $2.4908 \times 10^{-3}$ | 2.4908 | 2.5399 | $2.5399 \times 10^{-3}$ | 2.540 | $2.540 \times 10^{-2}$ | 1.8683 |
| 2.9886 | $2.9886 \times 10^{-2}$ | 29.886 | 30.479 | $3.048 \times 10^{-2}$ | 30.480 | 0.3048 | 22.419 |
| $0.2953 \times 10^{-2}$ | 3.3865 | 33.865 | 34.532 | $3.4532 \times 10^{-2}$ | 34.532 | 0.3453 | 25.40 |
| $1.0 \times 10^{-3}$ | $1.0 \times 10^{-5}$ | $1.0 \times 10^{-3}$ | $1.019 \times 10^{-2}$ | $1.019 \times 10^{-5}$ | $1.0198 \times 10^{-2}$ | $1.0198 \times 10^{-4}$ | $7.5006 \times 10^{-3}$ |
| 1.0 | $1.0 \times 10^{-2}$ | 20.0 | 10.194 | $1.019 \times 10^{-2}$ | 10.019 | 0.1019 | 7.5006 |
| $1.0 \times 10^2$ | 1.0 | 1.0 | $1.0197 \times 10^3$ | 1.019 | $1.0198 \times 10^3$ | 10.198 | $7.5006 \times 10^2$ |
| 0.10 | $1.0 \times 10^{-3}$ | 1.0 | 1.0197 | $1.0197 \times 10^{-3}$ | 0.10198 | $1.0198 \times 10^{-2}$ | 0.750 |
| $9.8097 \times 10^{-2}$ | $9.8069 \times 10^{-4}$ | 0.9807 | 1.0 | $1.0 \times 10^{-3}$ | 1.0 | $1.0 \times 10^{-2}$ | 0.7356 |
| 98.097 | 0.9807 | $9.8069 \times 10^2$ | $1.0 \times 10^3$ | 1.0 | $1.0 \times 10^3$ | 10.0 | $7.356 \times 10^2$ |
| $9.806 \times 10^{-2}$ | $9.8063 \times 10^{-4}$ | 9.806 | 0.9999 | $9.999 \times 10^{-4}$ | 1.0 | $1.0 \times 10^{-2}$ | 0.7356 |
| 9.8063 | $9.8063 \times 10^{-2}$ | 98.063 | 99.997 | $9.999 \times 10^{-2}$ | $1.0 \times 10^2$ | 1.0 | 73.554 |
| 0.1333 | $1.3332 \times 10^{-3}$ | 1.3332 | 1.3595 | $1.3595 \times 10^{-3}$ | 1.3595 | $1.3595 \times 10^{-2}$ | 1.0 |

## 2.5.4  Water Depth

The same equation as used for head pressure can be solved to find column height:

$$P = D \times H$$

$$H = P/D$$

where $P$ = measured pressure

$D$ = density of the liquid

$H$ = height of liquid

The same principle is used to measure the amount of liquid in a container or depth in a body of water. If the pressure is measured at the bottom of a container and the density of the liquid is known, the equation can be solved for H (height). Once the height is determined, the volume can be calculated.

Another application is a diver's depth gage. By measuring the pressure at an unknown depth and dividing by the density of the liquid (water), depth is determined. For example, 1 foot of water = 0.434 psi, the pressure reading is 10.84 psi:

$$H = P/D$$

$$H = 10.84/0.434$$

$$H = 25 \text{ ft}$$

This principle can be used to measure a few feet (wave heights) or depths hundreds of feet below the surface of the ocean.

## 2.5.5  Converting from One Unit to Another

In many instances, it may be necessary to convert from one pressure unit to another. If a pressure transducer is not specified for a specific application, it will be calibrated in pounds per square inch and to a standard pressure range of the manufacturer. Table 2.1 gives the conversion factors to convert from one pressure unit to another. To use the chart, locate the units you are converting from in the lefthand column and the units you are converting to in the horizontal column. Read the conversion factor at the intersection of the two columns and multiply the conversion factor times the number of units in the lefthand column.

*Example*: Convert 100 inches of water to pounds per square inch.

1. Locate inches of water in the lefthand column of the chart.
2. Move across the chart to the column headed psi.
3. Read $3.6127 \times 10^{-2}$ in column 2.
4. Multiply 100 inches water by 0.036127 to get 3.6127 psi.

*Example*: Convert 80 psi to bars.

1. Locate pounds per square inch in the lefthand column of the chart.
2. Move across the chart to the column headed bar.
3. Read 0.06895 in column 8.
4. Multiply 80 psi by 0.06895 to get 5.516 bars.

# 3
# Pressure Terminology

A number of terms used to specify pressure sensors and transducers require some definition. Such terms include gage or differential pressure, maximum overpressure or burst pressure, and the four basic types of pressure measurements. Pressure is measured relative to a reference pressure; the difference in the four types of pressure measurements is the reference pressure.

## 3.1 ABSOLUTE PRESSURE

Absolute pressure is measured relative to a perfect vacuum and is usually expressed in pounds per square inch absolute (psia). To make an absolute sensor, a reference vacuum is sealed inside the sensor.

### 3.1.1 Absolute Monolithic Sensor

Absolute solid-state sensors are processed in wafer form. A wafer is a 4- or 5-inch-diameter slice of silicon 0.016 inch thick. Resistors are diffused or ion-implanted into the top of the silicon wafer, and the back side of the wafer is masked and etched to form a thin silicon diaphragm. The resistors are aligned on the diaphragm in a Wheatstone bridge configuration—two resistors in tension and two in compression. This wafer can be used to produce any one of the four types of sensors. The wafer is then sealed to a second

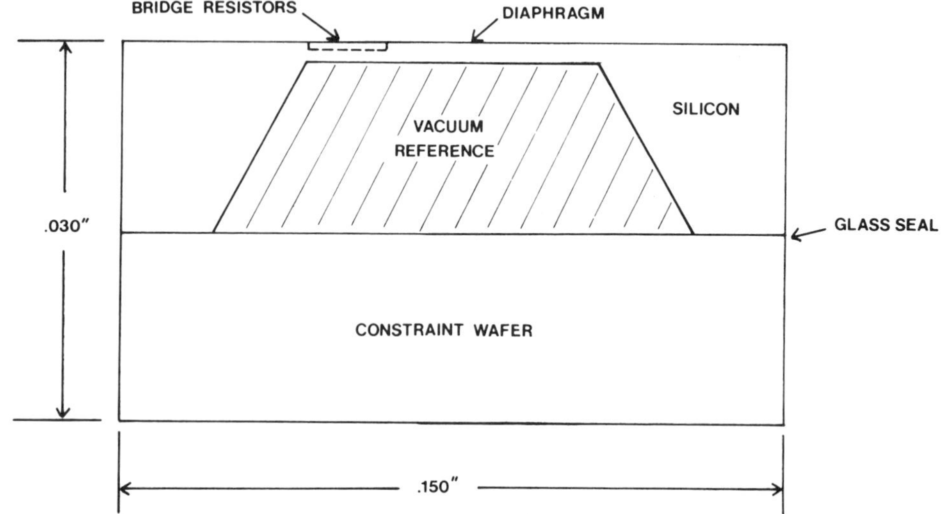


FIGURE 3.1   Solid-state absolute sensor.

wafer (sometimes called a *constraint wafer*) under a high vacuum.
The vacuum is now sealed inside the sensor (Figure 3.1). The
thin diaphragm area has a hard vacuum on one side and the unknown
pressure on the other side. If a hard vacuum is applied to the
sensor, the pressures are balanced and the output is zero. These
sensors are also called *vacuum sensors*.


## 3.2   GAGE PRESSURE

Gage pressure is measured relative to ambient atmospheric pressure
with a differential device that has the reference side of the diaphragm
vented to atmosphere.


### 3.2.1   Gage Monolithic Sensor

To produce a gage sensor, a hole is etched through the constraint
wafer to provide access to the back of the diaphragm (Figure 3.2).
One side of the diaphragm is vented to atmosphere; either side
of the diaphragm can be vented. To improve reliability, the top
of the sensor can be vented and the pressure applied to the back
of the diaphragm. Pure aluminum traces are used to make electrical
connections to the bridge resistors. Wires are bonded to these pads
during assembly of the sensor; hence they cannot be protected

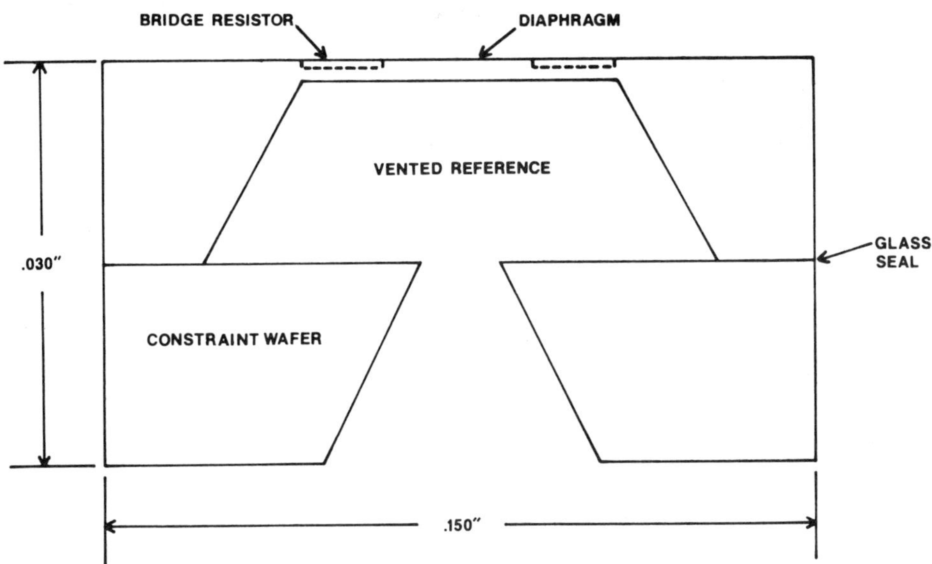


FIGURE 3.2   Solid-state gage sensor.

by covering with some inert material. On the back side, the medium
just comes in contact with the bulk silicon and the mounting surfaces.
The gage sensor will automatically compensate for changes in altitude
or barometric pressure. Gage sensors can be used to measure pressures less than atmospheric as well as positive pressures. Gage
sensors are used to measure liquid level, compressors, or engine
vacuum in a car.

### 3.2.2   Vacuum Gage

Vacuum gage is a term used with transducers more than with sensors.
It refers to a transducer that has been calibrated to measure pressures less than atmospheric. The term *vacuum* means any pressure
less than atmospheric—it is not possible to have a vacuum greater
than atmospheric pressure. The definition of a perfect vacuum is
the absence of all gaseous fluids.

## 3.3   DIFFERENTIAL PRESSURE

Differential pressure measures one input pressure relative to another.
Gage is a specific differential pressure in that one pressure is
always atmospheric.

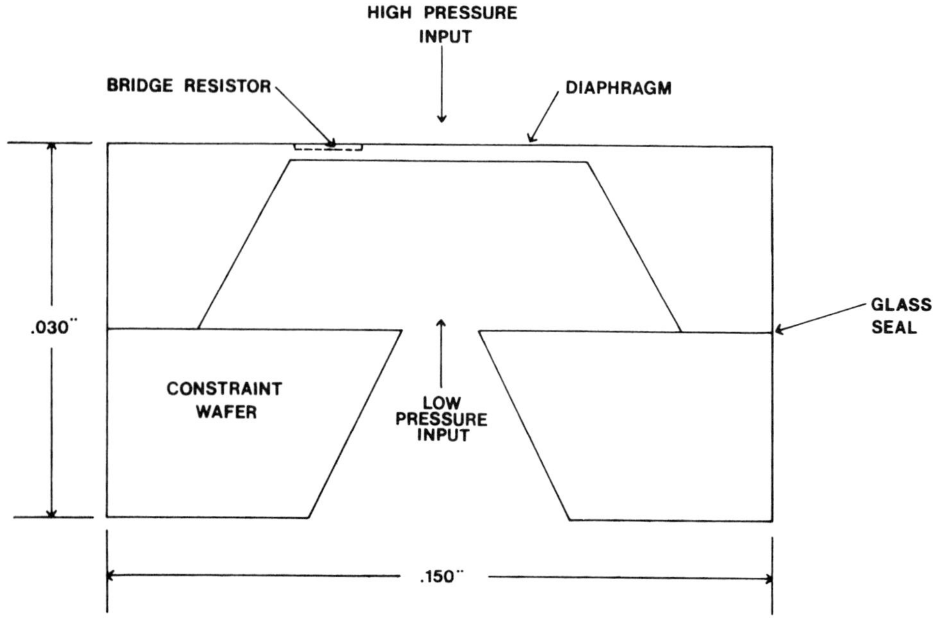


FIGURE 3.3   Solid-state differential sensor.

### 3.3.1   Differential Monolithic Sensor

The differential sensor is constructed exactly as is the gage sensor
(Figure 3.3). The only noticeable difference is that the differential
package has two pressure ports, while the gage has one port and
a vent. The differential sensor has one specification that the absolute
and gage sensors do not: maximum working pressure. Maximum
working pressure is the maximum pressure that can be applied
to either port. Since the most common application for differential
sensors is flow measurements, it is common to have pressures as
high as 1000 psi in common mode with 5-psi differential. The pressure
that is measured is the 5-psi difference, but the sensor has to
stand the 1000-psi working pressure. This makes the construction
of the differential package more complicated than that of the absolute
or gage sensor. Some differential transmitters are rated for 1000-psi
working pressure and 1- or 2-psi differential.

## 3.4   SEALED REFERENCE PRESSURE

If an absolute sensor is fabricated and a reference pressure other
than vacuum is used, it is referred to as a *sealed reference sensor*.

The most common reference is a standard atmosphere, but virtually
any pressure can be used as the reference. The pressure that
results in zero strain across the diaphragm can be controlled in
this manner. Sealed reference sensors can be purchased, or a differential sensor can be used by sealing one port. A sealed reference
sensor will react to pressure in the same way that an absolute sensor
does.


## 3.5  ATMOSPHERIC PRESSURE

Atmospheric pressure is the weight of the air surrounding the earth.
The average pressure at sea level is 14.696 psi, or 29.921 inches
of mercury. Air is a mixture of gases:

| Dry air | Percent by volume |
|---|---|
| Nitrogen | 78 |
| Oxygen | 21 |
| Argon | 0.93 |
| Other gases | 0.07 |

An other factor that influences air pressure is the amount of
moisture (humidity) in the air. Warm air will hold more moisture
than will cold air. When warm, wet air is cooled, clouds are formed
and rain or snow may follow. The effects of air pressure are generally overlooked because air is not visible. A straw could not draw
liquid from a glass without air pressure. The pressure on the top
of the liquid forces it up the straw. Altitude above sea level has
the largest effect on atmospheric pressure.


### 3.5.1  Altitude

The air surrounding the earth resembles an acean many miles deep.
The air is in constant motion, similar to the tides in the sea. The
sun heats the land masses, which heat the air and cause it to rise
and be replaced by cooler air. The atmosphere is divided into four
layers: troposphere, stratosphere, mesosphere, and thermosphere
(Figure 3.4). The troposphere extends 16 kilometers (10 miles)
above the surface of the earth; storms and clouds occur in this
layer. The stratosphere extends from 16 to 48 kilometers (30 miles);
het planes fly in the stratosphere. The mesosphere extends from
48 to 80 kilometers (50 miles) and has the lowest temperatures found
in the atmosphere. The thermosphere, above 80 kilometers, contains

just 1 percent of the air found in the atmosphere. Pressure decreases with altitude as follows:

| Altitude | Pressure | | Temperature |
|---|---|---|---|
| (ft) | (psi) | (kPa) | (°C) |
| 0 | 14.70 | 101.3 | 15.0 |
| 5,000 | 12.23 | 84.3 | 5.1 |
| 10,000 | 10.10 | 69.6 | -4.8 |
| 20,000 | 6.75 | 46.5 | -24.6 |
| 30,000 | 4.36 | 30.1 | -44.4 |
| 50,000 | 1.69 | 11.6 | -55.0 |
| 80,000 | 0.43 | 3.0 | -55.0 |

Change in pressure as a function of altitude is not linear. The change in pressure from 0 to 5000 ft is 2.47, compared to 2.13 from 5000 to 10,000 ft. Barometric pressure has an effect on atmospheric pressure, but not as large an effect as altitude.

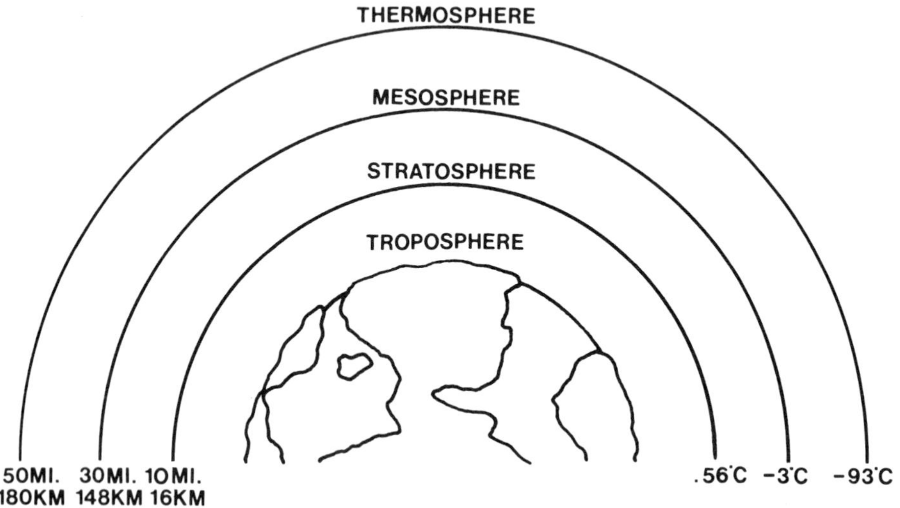


FIGURE 3.4   Four layers of the atmosphere.

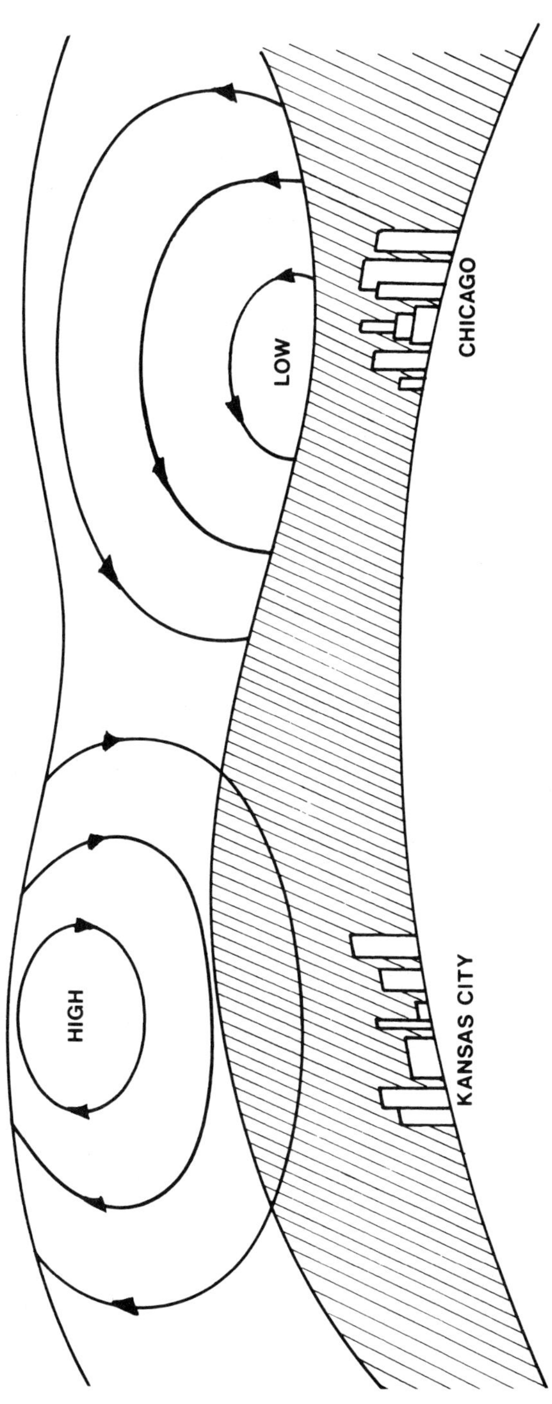


FIGURE 3.5 High- and low-pressure areas.

### 3.5.2  Barometric Changes

Atmospheric pressure varies from place to place. The ocean of air
is not uniform, but has valleys and hills (Figure 3.5). A high-
pressure area resembles a mountain of air; the winds spiral outward
and downward, causing warming and the evaporation of clouds.
Highs usually bring fair weather. A low-pressure area resembles
a valley; the winds spiral inward and force the air in the center
to rise. As the air rises, it is cooled and may condense to form
clouds and rain. A low-pressure area may bring wet, stormy weather.
A decrease in barometric pressure of 0.3 inch of mercury (0.147 psi)
in a matter of hours usually means that a front is passing through
and stormy weather will follow.

## 3.6  DEFINITION OF TERMS

The following are terms used to describe or characterize transducers,
sensors, or transmitters.

### 3.6.1  Parameters

*Reference Conditions*:  Specifies the conditions under which the
    device is tested: power-supply voltage, ambient temperature
    (usually 25°C), common-mode line pressure (working pressure)
    for differential devices, and temperature range for which the
    device has been compensated.
*Maximum Ratings*:  Gives the maximum conditions under which the
    device is specified to operate: power-supply voltage, maximum
    operating and storage temperature range, and the maximum
    pressure than can be applied to the device without damage.
*Operating Pressure*:  The pressure range for which the device
    has been calibrated and the parameters are guaranteed.
*Span*:  The device output signal over the pressure range (full-scale
    output - offset).
*Sensitivity*:  The change in output signal created by a specified
    change in input pressure.
*Reference Pressure*:  Pressure used as a reference in measuring
    device errors (0 psig = gage, 0 psid = differential, 0 psia =
    absolute).
*Offset Voltage*:  Device output signal when reference pressure is
    applied.
*Common Mode*:  Maximum pressure that can be applied to both input
    ports of a differential device (working pressure).

*Burst Pressure*: Maximum pressure that can be applied to a device without rupture of the sensing element or case.

### 3.6.2 Offset Error Terms

*Offset Calibration*: Maximum error band at reference conditions and reference pressure ($V_{offset} \pm V_{cal}$).

*Offset Repeatability*: The ability of the device to reproduce the offset voltage after exposure to any temperature or pressure in the specified range.

*Offset Stability*: The error band expressing the ability of the device to maintain the offset voltage at the reference conditions over a period of time (usually 1 year).

*Offset Temperature Coefficient*: The error band defined by the maximum deviation of the offset voltage as the temperature is varied from 25°C to any other temperature in the specified range.

*Offset Error*: The error band defined by the maximum deviation of the offset voltage from its specified value. The sum of calibration, repeatability, stability, and temperature errors.

### 3.6.3 Span Error Terms

*Span Calibration*: The error band defined by the maximum error in calibrating sensitivity ($V/psi \pm V_{cal}$).

*Span Repeatability*: The error band expressing the ability of the device to reproduce its span, measured at reference conditions, after exposure to any pressure or temperature within the specified range.

*Span Stability*: The error band expressing the ability of the device to maintain the span voltage at the reference conditions over a period of time (typically 1 year).

*Span Temperature Coefficient*: The error band defined by the maximum deviation of span as temperature is varied from 25°C to any temperature within the specified range. Expressed as $V/psi \pm V_{error}$.

*Hysteresis*: The maximum deviation in output signal when a specific pressure point is approached, first with an increasing pressure, then with a decreasing pressure, at a constant temperature.

*Linearity*: The closeness to which a curve approximates a straight line. As a performance specification, linearity is expressed as independent linearity, terminal-based linearity, or zero-based linearity. Independent linearity is assumed if expressed as linearity.

> *Independent*:  A line midway between two parallel straight lines
> closest together and enclosing all points on the calibration
> curve (Figure 3.6).
> *Terminal-Based*:  The maximum deviation of the calibration curve
> from a straight line intersecting the calibration curve at
> zero and full-scale values (Figure 3.7).
> *Zero-Based*:  The maximum deviation of the calibration curve
> from a straight line intersecting the calibration curve at
> zero value and minimizing the maximum deviation (Figure 3.8).

Independent linearity is specified only because it results in a
smaller number. In most applications, terminal-based linearity
is used; that is, offset is adjusted first, then full-scale pressure
is applied and span is adjusted.

*Span Error*:  The maximum deviation of span from its specified value,
including calibration, temperature error, linearity, hysteresis,
repeatability, and stability.

## 3.6.4 Overall Accuracy

*Accuracy*:  The sum of linearity and repeatability, or how accurately
the device will measure the input pressure.

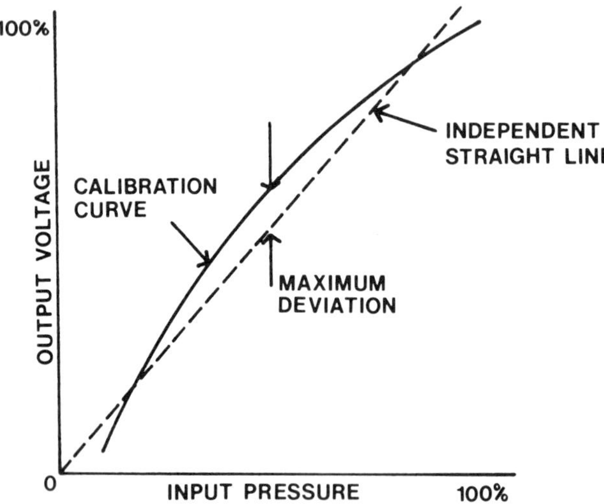


FIGURE 3.6   Independent linearity.

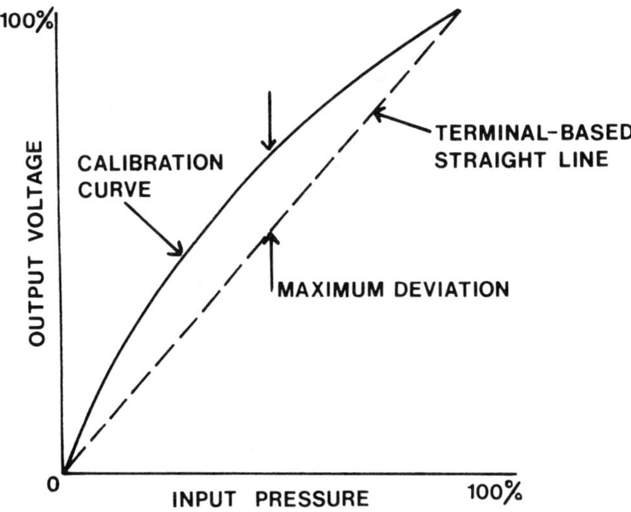


FIGURE 3.7   Terminal-based linearity.

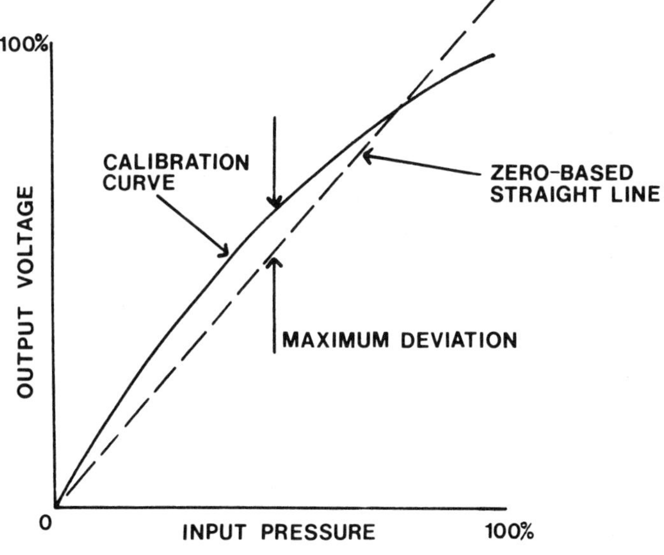


FIGURE 3.8   Zero-based linearity.

*Overall Accuracy*:   The combined error band relative to best straight
   line with forced reference. It excludes offset and span calibration
   error, but includes all other offset and span errors.
*Static Error*:   The error at reference conditions, including linearity,
   hysteresis, and repeatability. Does not include temperature
   effects. Used in transmitter specifications more than in sensor
   or transducer specifications.
*Dynamic Error*:   The error at reference conditions, including combined effects of linearity, hysteresis, and repeatability over
   the specified temperature range. Also used with transmitters.


### 3.6.5 Transmitter Terms

A few terms apply only to transmitters:

*Damping*:   Some transmitters have adjustable time constants from
   0 to 30 seconds.
*Turndown Ratio*:   The ability to change the sensitivity of a transmitter with an internal adjustment. Turndowns of 10 to 1 is
   common.


### 3.6.6 General Terms

*Analog Output*:   A voltage or current that is a continuous function
   of the measured parameter.
*Compensation*:   The addition of specific components to compensate
   a known error.
*Dead Volume*:   The volume of the input pressure port of a trans-
   ducer or sensor at ambient temperature and pressure.
*Full Bridge*:   A Wheatstone bridge with four active elements or gages.
*Half Bridge*:   A Wheatstone bridge with two active elements.
*Input Impedance*:   The impedance measured across the excitation
   terminals of the device.
*Load Impedance*:   The impedance presented to the output terminals
   of a device by the external circuitry.
*Cycle Life*:   The minimum number of pressure cycles a device can
   endure and remain within a specified tolerance.
*Strain*:   The ratio of the change in length to the initial unstrained
   reference length.
*Thermal Coefficient of Resistance*:   The change in resistance of a
   semiconductor resistor per unit change in temeprature.
*Stroke*:   The change in volume as pressure is applied.

# 4

# Common Transduction Methods

## 4.1 INTRODUCTION

One transducer design will not be suitable for all applications.
In a piece of portable equipment that has no power source (for
example, a fire extinguisher), a Bourdon-tube meter makes sense.
It requires no power and gives an indication of the charge in the
fire extinguisher. However, a sensor that has to interface with
a microprocessor requires an electrical output such as a piezoresistive
or capacitive device. In this chapter, we will discuss a few of the
more common transduction methods that have electronic outputs.

## 4.2 PIEZORESISTIVE SENSORS

Piezoresistive sensors (strain-gage sensors) are fabricated using
silicon processing techniques common in the semiconductor industry.
For this reason, they have taken on some of the semiconductor
terminology. Piezoresistive sensors are referred to as IC (integrated
circuit) sensors, solid-state sensors, monolithic sensors (formed
from single-crystal silicon), or just silicon sensors. The sensors
are processed in wafer form; each wafer contains 300 to 500 sensors,
depending on the size of the sensor die. A number of wafers can
be processed at one time, producing large numbers of sensor dies
per run.

### 4.2.1  Sensor Fabrication

A number of strain-sensitive resistors are ion implanted or diffused into silicon wafers using standard semiconductor techniques. These resistors are connected in a Wheatstone bridge configuration; two resistors increase with positive pressure, while the other two decrease with positive pressure. Alignment of the resistor on the silicon determines if the resistor will increase or decrease with applied pressure. Next, a diaphragm is created by etching or grinding the silicon from the back side. The diaphragm thickness determines the pressure range of the sensor. This ratio is not a linear function: Doubling the thickness decreases the sensitivity by a factor of four. As pressure is applied to the sensor, the resistors are strained, causing an imbalance in the Wheatstone bridge proportional to the applied pressure. The differential output of the sensor is then conditioned to interface with other electronic devices, such as a computer or voltmeter. Piezoresistive sensors are small and can be manufactured in high volumes to reduce the selling price.

## 4.3  LINEAR VARIABLE DIFFERENTIAL TRANSFORMERS (LVDTs)

LVDT stands for linear variable differential transformer. LVDTs are used to convert small position changes into electrical signals. LVDT sensors are used for position sensors, accelerometers, and pressure sensors. There are three types of transformers: the LVDT measures linear displacement, the rotary differential variable transformer (RVDT) measures angular displacement, and the linear phase differential transformer (LPDT) measures the phase of the output with respect to the phase of the transformer primary. We will limit this discussion to the LVDT transformer used as a pressure sensor.

### 4.3.1  LVDT Transformers

The transformer has two secondary windings and a movable core (Figure 4.1). The LVDT converts small position changes into electrical signals. As the core moves, the signal induced in one secondary winding increases and the signal induced into the other secondary winding decreases. These two secondary windings are wound in such a manner that their output voltages oppose or cancel. When the core is positioned at the center (null position), the output is zero (winding one = winding two). When the core is moved through the null position, the output undergoes an abrupt phase reversal.

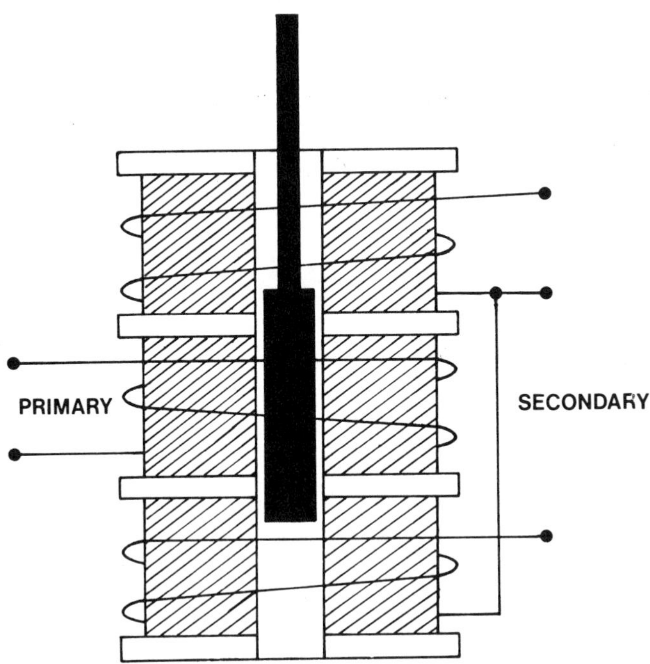



FIGURE 4.1  Linear variable differential transformer.

The phase of the output indicates on which side of center the core is positioned, and the magnitude indicates the distance from the center of the core (see Figure 4.2). Over a specified core travel, there is a linear relationship between LVDT output and core displacement.

## 4.3.2  LVDT Transducers

To use an LVDT as a pressure transducer, the pressure must be converted to linear displacement. Any pressure-sensitive device that changes size or shape as pressure is applied—such as bellows, Bourdon tubes, or aneroid capsules—can be used. The bellows or Bourdon tube is attached to the movable core of the transformer and, as pressure is applied, the core is displaced from the null position, causing an output voltage. The transformer is a differential device and can be used to make gage, absolute, or differential transducers.

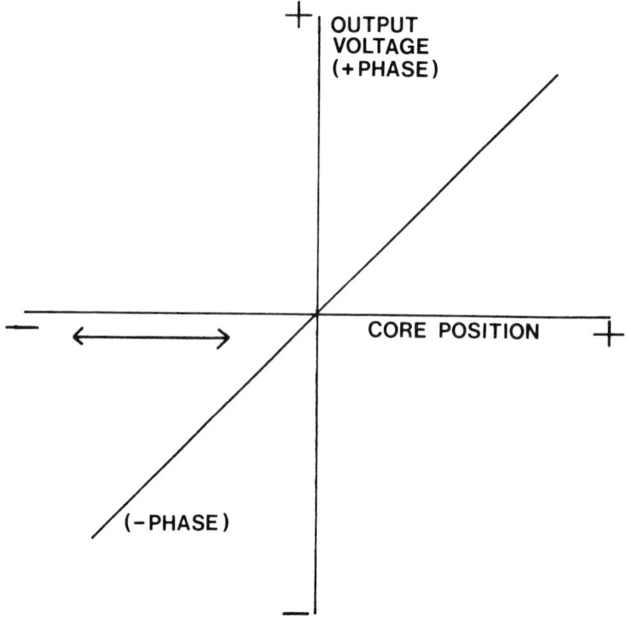


FIGURE 4.2   LVDT displacement response.


## 4.3.3  Signal Conditioning

The LVDT device also needs signal-conditioning electronics. These
circuits provide AC excitation voltage for the transformer, an amplifier stage, and a demodulator to convert the AC output to a DC
(direct current) voltage (see Figure 4.3). The LVDT output is
proportional to the excitation voltage, and the excitation voltage
generator must have a stable amplitude and frequency. The amplifier
and AGC (automatic gain control) circuit amplify the excitation
voltage oscillator and provide a stable amplitude drive voltage to
the LVDT. The amplifier also provides an AC signal to the demodulator. Typically, synchronous demodulators are used rather than
rectifier demodulators, because they are sensitive to phase and
are easy to implement. Maximum linearity occurs when the phase
of the LVDT output is exactly synchronized with the demodulator
reference signal. Phase shift between the primary and secondary
voltages of the LVDT is common. Figure 4.3 shows a phase-shift
network used between the excitation amplifier and demodulator to
adjust the phase of the reference signal to that of the demodulator.
This RC (resistance-capacitance) network is adjusted during calibration to minimize linearity errors due to stray capacitance.

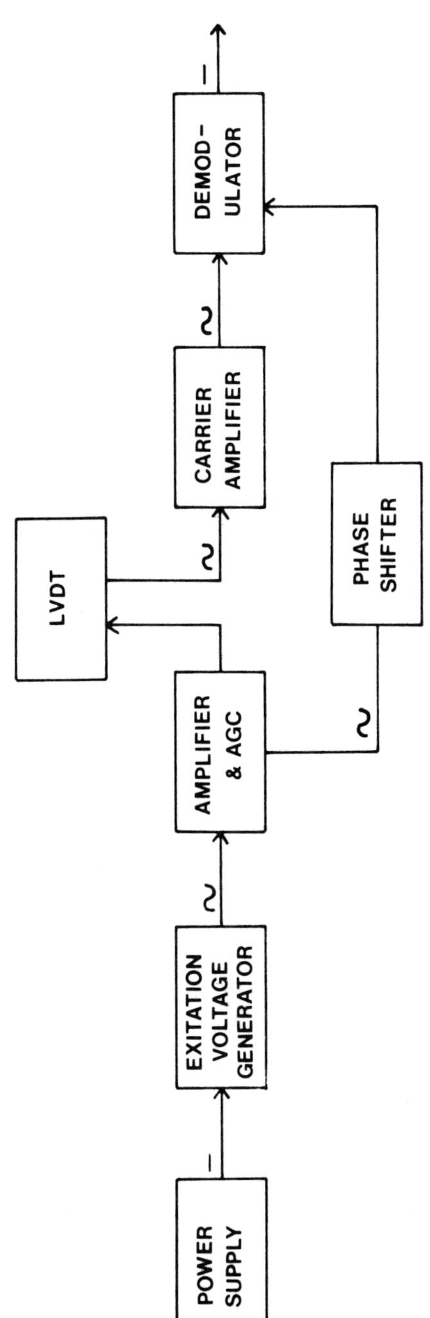


FIGURE 4.3  LVDT signal-conditioning.

## 4.4  CAPACITIVE TRANSDUCERS

Over half of the transmitters installed in the chemical industry
are capacitive devices. These transmitters are available in absolute,
gage, and differential pressures and range from 1 inch of water
full scale to as high as 5000 psi. Accuracies of 0.1 to 0.5 percent
of full scale are common for this type of transmitter. The differential
transmitter is generally used to measure flow and may have to measure differential pressures as low as a few inches of water, with a
working pressure as high as 1000 psi.


### 4.4.1  Capacitive Cell

The heart of the capacitive sensor is the capacitive cell, shown
in Figure 4.4. The capacitive cell consists of two cell halfs, each
with a fixed capacitor plate, a flexible center diaphragm that senses
the pressure variations, and two isolating diaphragms that are back-
filled with oil to the sensing diaphragm. The capacity is a function
of the size of the capacitor plates, the dielectric constant of the
insulating material, and the distance between plates. The capacity
can be calculated as follows:

$$C = KA/4\pi d$$

where  C = capacitance

      K = dielectric constant of insulator

      d = distance between plates

      A = area of plates

Pressure applied to the cell deflects the diaphragm and creates
an increase in the capacity to one plate (decreasing the distance
between plates), while decreasing the capacity to the other plate
(increasing the distance between plates). From the equation, if
the area of the capacitor plate is a constant and the dielectric constant of the insulator is a constant (oil-filled or a vacuum), changing
the distance between plates will cause a linear change in the capacitance. By designing a signal-conditioning circuit that is sensitive
to the capacitance ratio but not to the absolute capacitance value,
the cell can be made less sensitive to temperature and working-
pressure variations. The diaphragm and stationary plates are constructed from metal or metal-coated quartz materials. The capacitive
plates and sensing diaphragm must be electrically isolated from the
outer package.

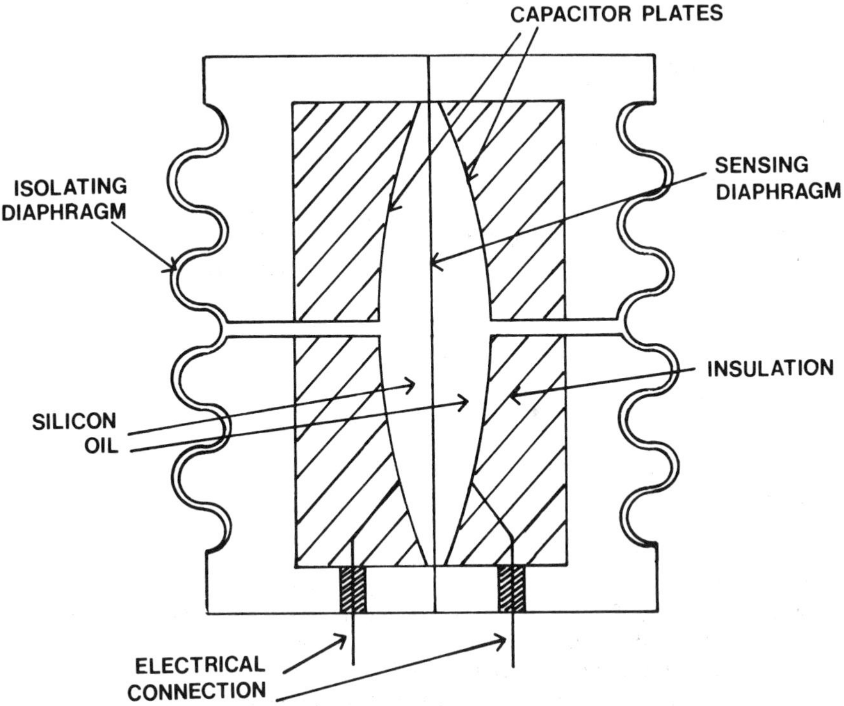


FIGURE 4.4  Capacitive cell.

## 4.4.2  Cell Construction

The mechanical construction of the cell is very important to the
overall performance of the transmitter. If the package should expand
due to applied pressure, the position of the capacitor plates would
change, causing an error in the sensor response curve. The low-
pressure differential flow sensor would be the most critical. It is
common to measure a few inches of water differential pressure while
working with a system pressure as high as 1000 psi. The sensing
diaphragm has the same pressure on both sides and reacts just
to the difference of the two pressures, while the package has to
withstand the total working pressure and not change shape or position due to the forces caused by the high working pressure.

## 4.4.3  Silicon Capacitive Pressure Sensors

Sensor cells are also fabricated from silicon, using the same basic
design as the metal device. Single-crystal silicon sensors have

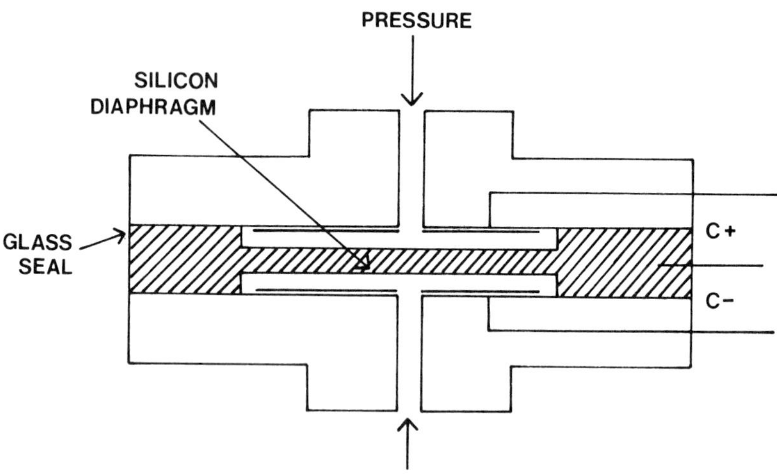


FIGURE 4.5   Silicon capacitive cell.


excellent elasticity and virtually no hysteresis. The sensing diaphragm
is micromachined from silicon, creating a diaphragm that is held
rigidly at the edges and flexes at the center (see Figure 4.5).
The sensing diaphragm is sealed between two rigid plates to complete
the basic capacitor cell. Silicon is generally sealed with low-melting-
point glass to give a hermetic seal, instead of being welded, as
is the metallic cell. Like the metallic cell, the silicon unit is available
in absolute, gage, and differential configurations.


## 4.5   STRAIN-GAGE SENSORS

As the name implies, bonded strain-gage sensors are sensors that
are fabricated by bonding a strain-sensitive resistor to a metallic
diaphragm. The technique used to bond the sensor to the diaphragm
is critical: It must hold the resistor to the diaphragm without changing characteristics due to aging, work-hardening (constantly being
flexed by the applied pressure), or creeping (not returning to
the original shape after being pressure cycled). Bonded strain-gage
sensors have a built-in advantage, in that they do not have to be
isolated from the working medium. Since there must be a diaphragm
to bond the strain-gage resistors, the diaphragm also isolates the
resistors from the working medium. These diaphragms are generally
stainless steel or some other resistant metal. Both semiconductor
and metal-foil resistors are used as sensing elements. The diaphragm
dimensions (diameter and thickness) determine the pressure range

of the transducer. This type of sensor can be designed to measure very high pressures, in excess of 10,000 psi. The very-high-pressure devices have small-diameter pressure passages to minimize the force the package has to withstand. The low- and mid-range sensors have larger diaphragms and larger diameter pressure ports. The sensor housing is generally made from stainless steel, and the entire assembly is welded together to give a strong, medium-resistant sensor (see Figure 4.6).

## 4.5.1 Operating Principle

The resistors are arranged in a Wheatstone bridge configuration to improve sensitivity. The strain-sensitive resistors can be connected

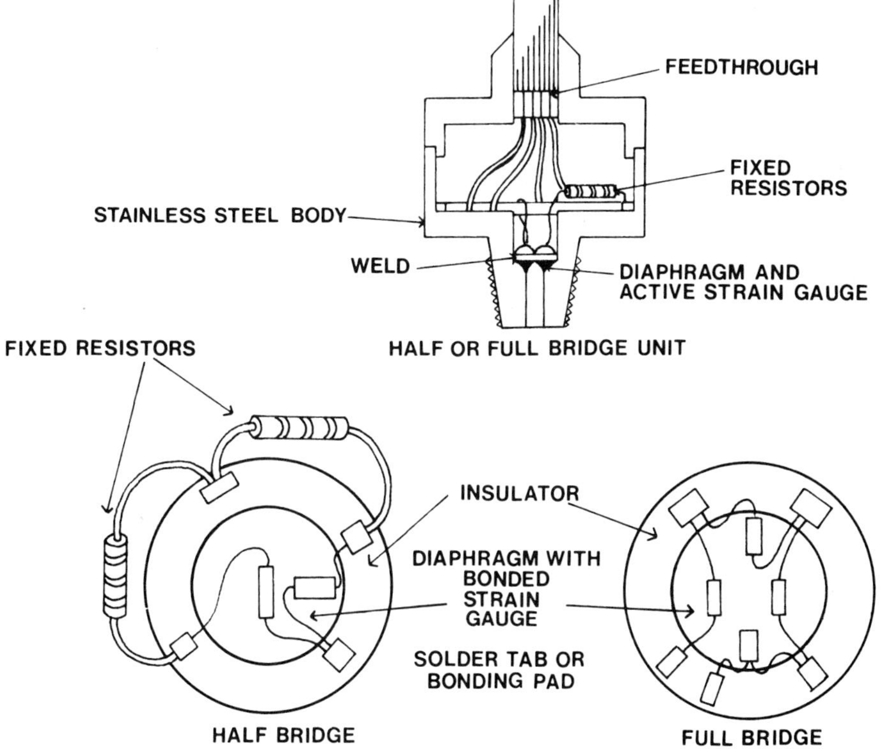


FIGURE 4.6 Sensor construction.

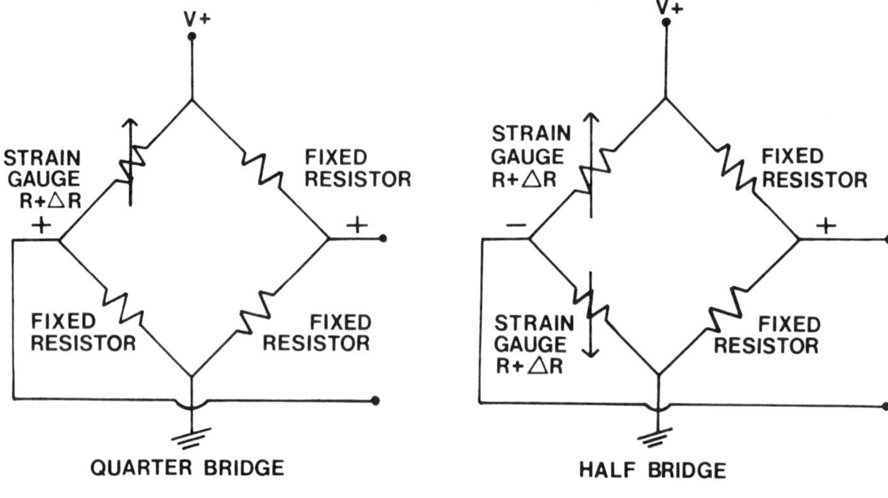



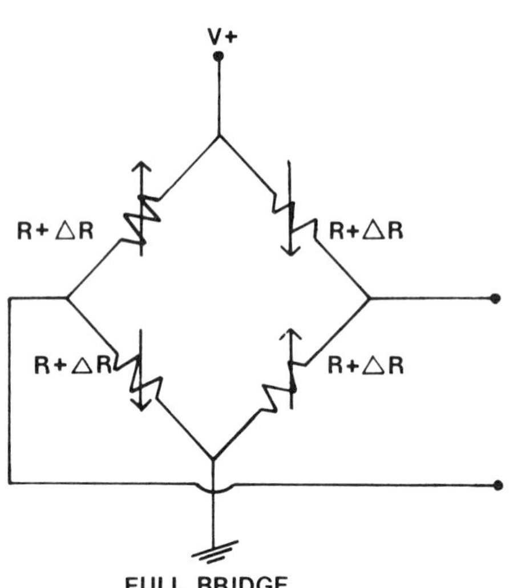



FIGURE 4.7   Bridge configurations.

in quarter-bridge, half-bridge, or full Wheatstone bridge configuration (see Figure 4.7). The quarter bridge uses three fixed resistors and the half bridge uses two fixed resistors to complete the Wheatstone bridge. These resistors are not strain sensitive and are used for bias purposes only. The half- and full-bridge configurations are the more common configurations. The resistors on each side of the bridge should have the same temperature coefficient to minimize the offset temperature coefficient of the bridge. The two strain-sensitive resistors are aligned so that one is in compression and the other in tension. The resistor in tension increases with applied pressure, while the resistor in compression decreases, unbalancing the bridge. The sensor will measure both positive and negative pressures. When a negative pressure is applied to the sensor, the diaphragm is strained in the opposite direction, and the resistor that was in compression goes into tension, while the resistor that was in tension goes into compression. This results in a sign reversal at the output. When no pressure is applied to the sensor, the bridge is balanced and the output is 0. When a positive pressure is applied to the sensor, the differential output increases (+mV) or, if a vacuum is applied, the differential output decreases (-mV).

## 4.6  PIEZOELECTRIC SENSORS

When properly cut and oriented with respect to its crystallographic axis, a piece of quartz or polycrystalline ceramic will generate a small electric charge when strained. Piezoelectric sensors require no external power and are used for dynamic measurements. These sensors have very high natural frequencies and low noise. Most accelerometers are piezoelectric because of their high-frequency response. Piezoelectric accelerometers are available with frequency responses ranging from 1 Hz to 200 kHz. Piezoelectric pressure sensors are generally used to measure very fast pressure changes, such as shock waves from explosions or a very fast pressure spike. Another very common application for piezoelectric sensors is in microphones, due to their small size and wide frequency response.

### 4.6.1  Operating Principle

In the early designs of piezoelectric sensors, a solid piece of quartz was cut and two surfaces were metallized for electrical contact. This structure was then mounted in a metallic housing that preloaded the crystal and was used as the electrical ground for the circuit. The end of the sensor was flexible, to couple the pressure change

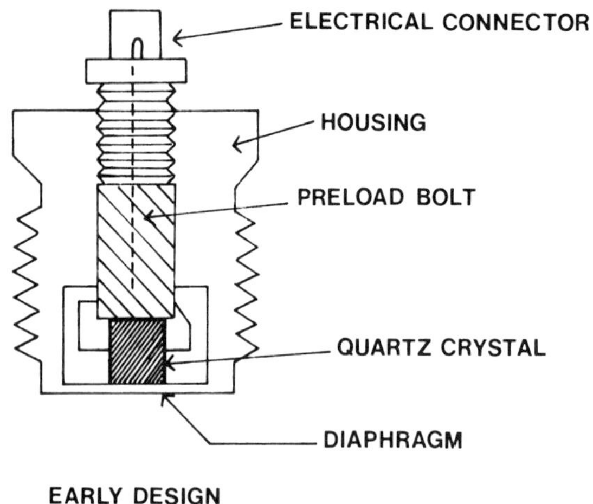



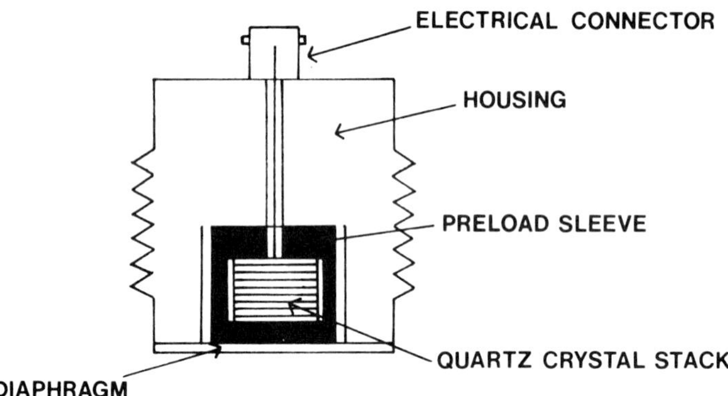



FIGURE 4.8   Piezoelectric sensor construction.

to the crystal. Today's sensors use a stack of thin crystal wafers to increase sensitivity. The crystals are stacked in series (negative surface of one crystal in contact with the positive surface of the next crystal) (see Figure 4.8). A single electric contact is made from the crystal through a connector with the housing acting as the signal return. As shown in Figure 4.8, pressure is applied to the diaphragm, which applies a compression force to the crystal element. As the crystal is compressed, a charge proportional to the pressure is generated.

### 4.6.2 Signal Conditioning

The output of the crystal is a charge, not a voltage. Care has
to be taken when signal-conditioning the output. Static charge
is related to the circuit capacity by

$$Q = CE$$

$$E = Q/C$$

where $Q$ = static charge

$\quad\quad C$ = capacity

$\quad\quad E$ = applied voltage

Stray capacitance can shunt the output signal to ground. The capacity of the cable from the sensor to the amplifier circuitry needs
to be considered when calibrating the system. The input impedance
of the amplifier circuitry is extremely important. If the impedance
is too low, the input capacity is shunted by a small resistor, resulting in large errors. Special amplifiers called *charge amplifiers* are
available to signal-condition the sensors. The charge amplifier has
very high input impedance and a range capacitor designed into
the input stage. It is not as sensitive to distributed capacity as
a conventional operational amplifier would be.

# 5

# Piezoresistive Sensors

## 5.1  INTRODUCTION

*Piezo* means pressure (a piezometer is an instrument for measuring
pressure) and *resistive* means the opposition to DC current flow.
Thus, a piezoresistive sensor is a device that changes resistance
as a function of applied pressure. Piezoresistive sensors are silicon-
based devices. Silicon is the second most common element on earth—
sand, quartz, agate, and opal are a few of the forms in which
the oxide appears. Silicon is a principal ingredient of glass and
an important ingredient of steel.

Piezoresistive sensors are manufactured from hyperpure silicon.
Single-crystal silicon is prepared by melting silicon in a controlled
atmosphere and drawing round ingots of pure silicon. These ingots
are then sawed into round wafers to be processed. Two- and three-
inch wafers used to be the standard sizes, and all the processing
equipment was designed for this size. Due to advances in the semiconductor industry, 4-, 5-, and 6-inch wafers are being processed
today. Obviously, the larger the wafer, the more devices can be
processed at a time—but the large investment in new capital equipment complicates the decision to change wafer size.

Silicon is an ideal choice for electromechanical devices. The
electrical properties are well known, due to silicon's wide use in
the semiconductor industry and, being a single-crystal structure,
the mechanical properties—such as strength, freedom from mechanical
hysteresis, and the ability to batch process—make low-cost, high-
performance pressure sensors possible. The following discussion
will deal with the design and fabrication of piezoresistive sensors.

## 5.2  SILICON PROCESSING

*Micromachining* refers to a process that enables precise three-dimensional shapes to be formed in silicon. Silicon micromachining is based on many of the same techniques used in integrated-circuit manufacture and is ideal for large-volume manufacturing of miniature devices requiring precise dimensional control. Presently, the technology allows control of dimensions from a few microns to a few millimeters, with tolerances of less than one micron. Micromachining has made possible structures such as diaphragms, bridges, cantilever beams, precision grooves, holes, and miniature probes. The process of micromachining involves various steps, such as photolithography (precise pattern transfer), three-dimensional silicon etching, and thin-film deposition.

### *5.2.1  Masking*

Photolithography is a process in which a light-sensitive photoresist is applied to the wafer and a photomask is used with ultraviolet light to transfer very precise patterns to the silicon wafer (see Figure 5.1). The entire wafer is covered with a masking layer, such as silicon dioxide or silicon nitride; a photoresist coating is added; the mask is placed over the wafer; and the resist is exposed to ultraviolet light. The exposed photoresist is then removed and the silicon is ready to be processed. A sensor will have a different mask for each processing step; there could be 8 to 12 masks used in the processing of one chip.

The desired pattern is drawn on a large scale to get very precise dimensions and then is reduced to the proper size. This pattern is then reproduced by a step-and-repeat process, until a mask containing hundreds of patterns and large enough to cover the wafer has been produced. There can be several masks for one processing step; for example, if certain transistors or resistors are desired on some, but not all, devices, one metal mask that connects all the devices can be generated and a modified mask that connects only selected devices can also be generated and used in place of the standard. Today, computers are used to generate masks for very complex devices. The computer can store the basic mask dimensions and make modifications as required.

### 5.2.2  Silicon Etching

Silicon-etching processes can be classified as *isotropic* or *anisotropic* and can be accomplished by either wet chemical or dry etching

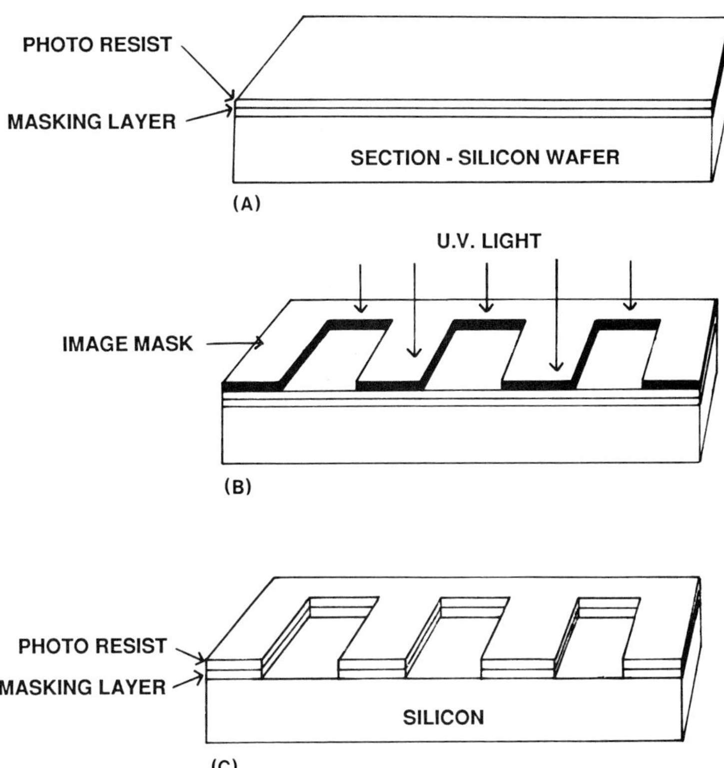



FIGURE 5.1  Photolithography.

techniques. Isotropic etching of silicon has no directional preference, and lateral dimensions cannot be easily controlled because of undercutting below the masking layer. Isotropic etching is not widely used for precision micromachining, but is used for the removal of bulk material.

Anisotropic chemical etching uses the fact that the etch rate is dependent on silicon's crystallographic orientation and impurity-doping density (see Figure 5.2). The etch rate in the (111) plane is much slower than that in any of the other planes, with some etchants as high as 400 to 1. The final etched shape will be bounded by the (111) plane, as shown in Figure 5.2.

Micromachining is used to create the diaphragm on piezoresistive sensors. Some of the key features needed are

1.  Ability to create a precise thickness
2.  Repeatable diaphragm thickness from run to run

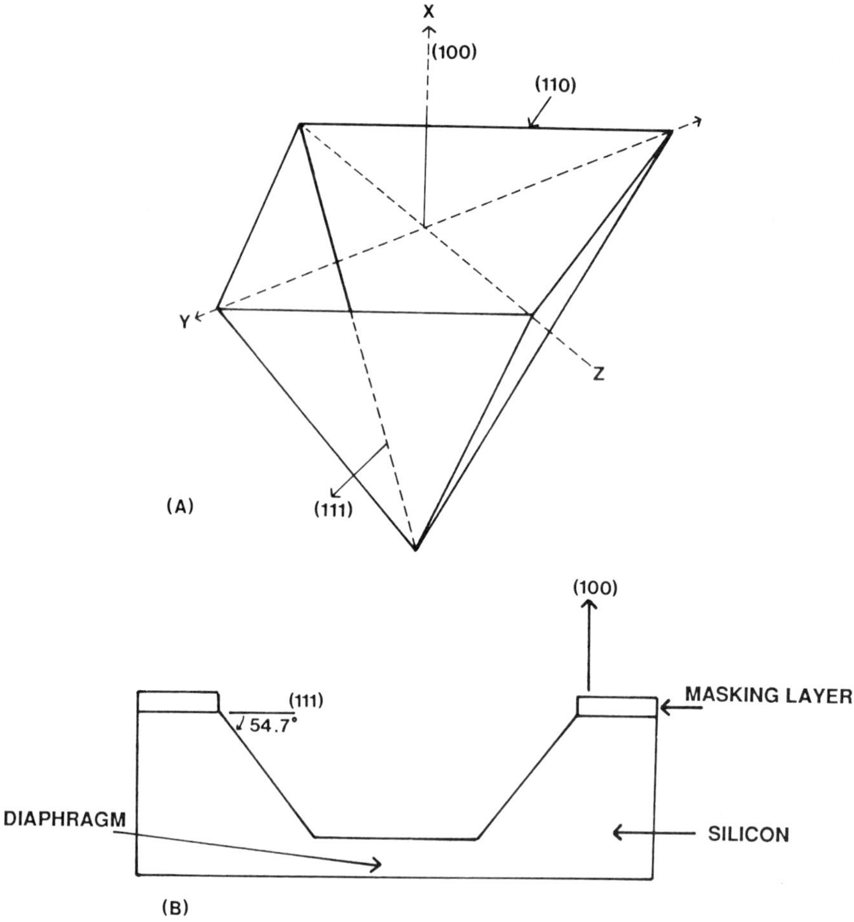



(A) THREE DIMENSIONAL VIEW OF (111) PLANE ON A (100)
SILICON WAFER.
(B) PATTERN ALIGNED TO THE (111) DIRECTION AND ATTACHED
ON A (100) WAFER.

FIGURE 5.2   Silicon etching.

3.   Uniform thickness across the wafer
4.   Smooth diaphragm free from irregularities
5.   Ability to vary the diaphragm thickness

The diaphragm is very important in making a piezoresistive sensor.
The sensitivity varies as the square of the diaphragm thickness.
Everything else being equal, a 20-micron diaphragm will have approximately 4 times the sensitivity of a 40-micron diaphragm. Also, if
the thickness should vary across the wafer, the sensitivity range
of that wafer could be 4 to 1 or more. One mask set can be used
to manufacture many pressure ranges by changing just the diaphragm
etch mask and etching a range of diaphragm thicknesses.

The diaphragm is created by masking the back side of the wafer;
the mask must be aligned to the crystallographic orientation of
the wafer. The wafer is then etched to the desired thickness. Small
diaphragm thicknesses can be controlled by P-stop techniques,
where a P+ region is diffused into the silicon or a slight etch is
done on the top side. These techniques are used for square and
rectangular diaphragms; round diaphragms have to be machined
into the silicon using different techniques.


## 5.2.3  Electrical Contacts

Electrical connections must be made to the Wheatstone bridge circuit
on the chip. This is generally accomplished by depositing metal
traces on the outer perimeter of the top side of the chip and welding
a gold or aluminum wire to the trace. Metal traces are not deposited
over the diaphragm region so as not to create a bimetal structure.
Low-value resistors are used to connect the elements in the diaphragm
region A metal mask is generated that provides metal traces from
the active components on the chip to metal bonding pads that are
used to connect the chip to the rest of the electronic circuit. These
bonding pads are generally pure aluminium, although more exotic
metals can be used. To make a gold bonding pad, two or three
metals are deposited on an aluminium base, ending with a gold surface. This is a very expensive process, and most manufacturers
choose to protect the aluminium by isolating the chip from the working medium. These bonding pads are very important, because they
determine what chemicals can come in contact with the die. To measure pressure, the medium must come in contact with the chip or
be coupled through an inert liquid. If pure aluminium bonding pads
are used, anything that oxidizes aluminium can cause a sensor failure
by attacking the aluminium and causing an open circuit. Since wire
bonding is one of the last process steps in the sensor assembly,

the pads cannot be coated until the assembly is complete. This
is a major problem with piezoresistive sensors. Most manufacturers
cover the die with a silicon-based gel that cures flexibly. The gel
stays the consistency of a soft gelatin. This prevents the coating
from generating unwanted stresses at the chip and, at the same
time, protects the bonding pads from high-humidity atmospheres
and liquids that might condense in the package. This coating will
not act as an isolator and will protect the chip from high-pressure
liquids.

## 5.3  SENSOR CHIP CONSTRUCTION AND PACKAGING

After the top wafer has been processed, it is ready to seal to a
silicon constraint wafer or to a Pyrex® plate, depending on the
cavity seal used. Then, the wafer is sawed into individual sensor
dies and packaged.

### 5.3.1  Reference Cavity Seals

Piezoresistive sensors are generally fabricated by sealing two wafers
together, creating a reference chamber. A top wafer that has the
bridge resistors ion implanted or diffused, metal traces deposited,
and a back-side etch that creates the diaphragm, is sealed to a
second wafer (sometimes called a constraint wafer). This results
in a chip with the active component on the top surface, a reference
cavity in the center, and a second wafer that may or may not have
a hole through it, depending on the type of sensor to be manufactured. Absolute sensors have a sealed reference cavity, while in
gage and differential sensors, the cavity is vented through the
constraint wafer to the atmosphere or a second pressure port. This
seal is very important to the operation of the sensor. If the seal
should leak, the offset voltage of the absolute sensor would change,
and the gage or differential pressure would have large errors.
   The two most popular sealing processes are low-temperature
glass (silicon to silicon) or electrostatic bonding to a Purex® constraint. This seal is generally done in a vacuum. To make a glass
seal, the constraint wafer is coated with a low-temperature glass
frit and heated to give a smooth layer of glass. The two wafers
are then placed in a fixture and reheated to reflow the glass. The
reflow process can be done in an evacuated bell jar, generating
a hard vacuum reference. If the process is done at ambient pressure,
a partial vacuum is generated due to the high heat. The result is
a hermetic seal across the entire wafer.

The electrostatic process requires a Pyrex  constraint; a seal is formed between the silicon top wafer and the Pyrex . The two wafers are fixtured, and a current passed through the wafers at an elevated temperature forms a molecular bond.

After the wafers are sealed together, the final wafer is sawed into individual dies to be assembled into pressure sensors.


## 5.3.2  Sensor Mount

A sensor that is designed to measure pressure will also be sensitive to stresses transmitted to the chip through the mount. Some sensor designs are more sensitive to mount stresses than others. The mount should have the following characteristics:

1.  The coefficient of thermal expansion should match the sensor material as closely as possible to minimize thermally induced strains.
2.  The mount should have no mechanical hysteresis.
3.  The mount should not creep or change with time.
4.  In gage and differential sensors, the mount must seal the pressure port to the back side of the chip.
5.  The mount must be compatible with the working medium.

Because the sensing die is extremely small (typically 50 to 150 mils square), the mount does not have to sustain large forces. On an absolute sensor, the mount just holds the die in place; on gage and differential devices, the mount is used as a pressure seal as well. If a pressure is applied to the back side of the diaphragm, the mount must sustain the following forces:

Assume a square diaphragm 80 mils on a side:

area = $0.80^2$ = 0.0064 square inch

An applied pressure of 100 psi results in a force as follows:

force = 100 × 0.0064 = 0.64 pounds

Several approaches are used to mount the sensor die. The first uses a compliant mount that cannot generate or transmit stresses to the die. The second tries to match the mount material to the sensor and package material. The first method is by far the more popular; most low- and medium-priced sensors have RTV mounts. RTV, being a soft, rubber-based material, will not couple strains from the package to the die. The main disadvantage of an RTV mount is its incompatibility with some chemicals. Most freons and gasoline will cause RTV to swell and become gummy.

Another mount used in piezoresistive sensors is a gold eutectic die. If gold and silicon are heated to just over 400°C, a eutectic is formed where the gold penetrates the silicon, creating a very strong bond. This mount is much stronger than RTV and does not have the medium-compatibility problems. The disadvantage of the eutectic mount is that it couples body stresses to the die and is not used on low-pressure sensors.

Another approach is to machine a structure out of silicon or Pyrex® that seals to the back side of the die and isolates the sensor from the package. This approach is much too expensive for the majority of sensors.


### 5.3.3  Sensor Package

Piezoresistive sensor packages can be divided into two groups: the printed-circuit-board (PCB) mount and the pipe mount.

The printed-circuit-board device is small and generally has barbed fittings for pressure connections and solder pins for electrical connections. The packages are generally plastic or glass-filled nylon with small mounting holes through the package. Printed circuit boards are mounted at standard spacings to conserve space, and the sensors have to conform to this spacing. PCB-mounted sensors are generally low pressure (below 100 psi) and measure clean air or gases.

The second package type is the pipe mount. This package is attached to the system through a threaded pipe connection and can measure pressures as high as 10,000 psi. These sensors are mounted outdoors and must endure the elements as well as any physical abuse.

These packages are not interchangeable: The PCB-mounted sensor will not survive outdoors, and the pipe mount will not fit on a printed circuit board.


## 5.4  JUNCTION ISOLATED PRESSURE SENSOR

Integrated piezoresistive sensors are monolithic silicon chips. The resistive element is defined by ion implanting boron into the N-type silicon substrate. This creates a P-type resistor isolated from the rest of the N-type silicon by the depletion region of a PN junction. This PN junction is back-biased by connecting the N-type silicon substrate to a positive potential. If a conductive medium is applied

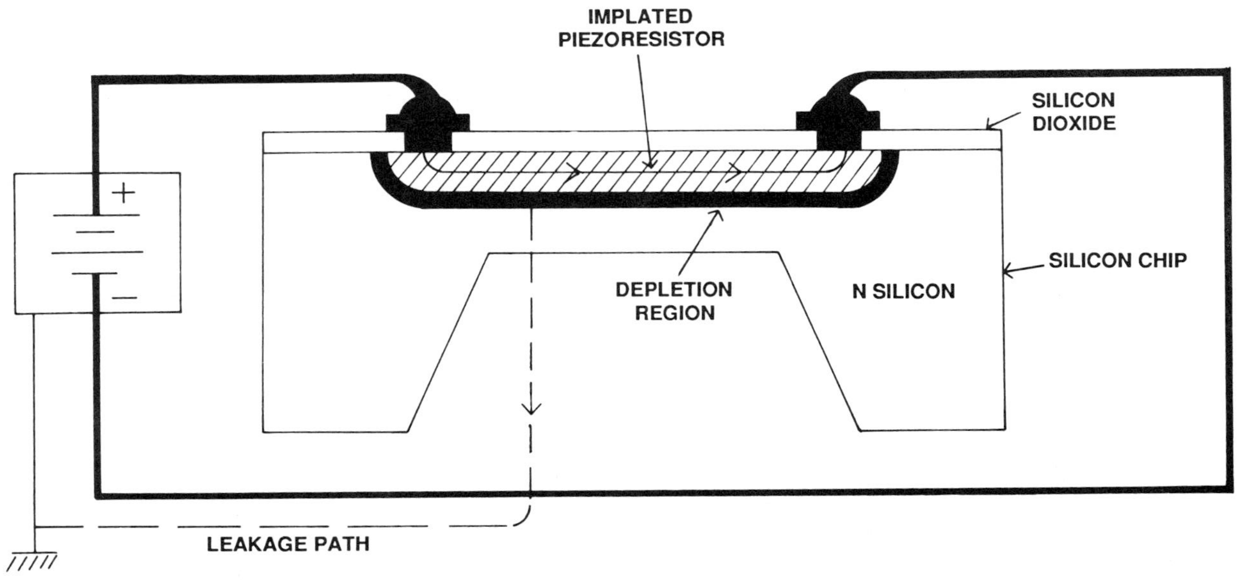


FIGURE 5.3   Junction–isolated sensor.

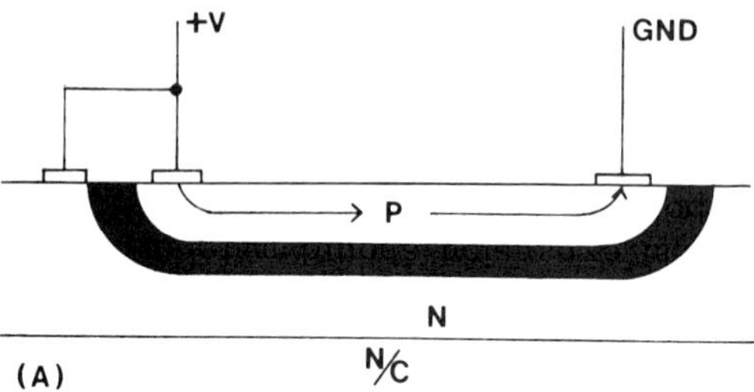



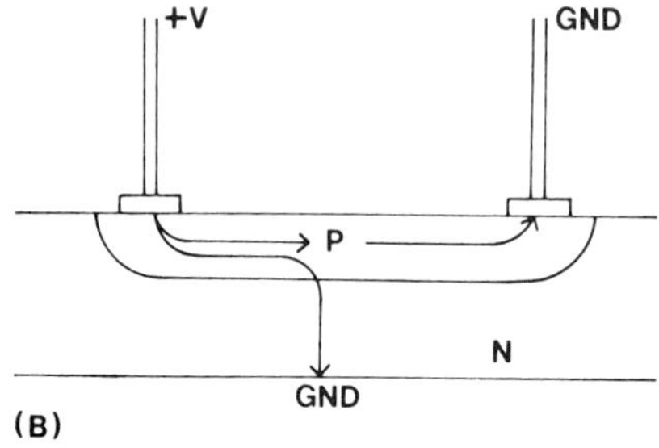



(A) CHIP SUBSTRATE TIED TO MOST POSITIVE POTENTIAL.
    (MIN. LEAKAGE)
(B) CHIP SUBSTRATE CONNECTED TO GND THROUGH THE
    MEDIA.

FIGURE 5.4   Biasing-PN junction.

to the back side of the chip and has an electrical path back to
the power-supply ground, a leakage path is established from the
chip to the supply (see Figure 5.3). A low-current leakage is always
present across a back-biased PN junction. With 1 mA of DC current
flowing through the resistor, the leakage current will be on the
order of 5 nA at room temperature (25°C) (see Figure 5.4). As
a rule, this leakage current doubles with every 8°C of temperature
increase. This means that a 5-nA leakage current at room temperature
will increase to 20 μA at 120°C and to 1.3 mA at 170°C (see Figure
5.5). This leakage current can cause a large error at elevated
temperatures. To minimize resistor leakage currents, the substrate
should be connected to the highest potential in the circuit and
care should be taken not to ground the back side of the chip through
the medium. This junction leakage also limits the maximum operating
temperature of silicon sensors.

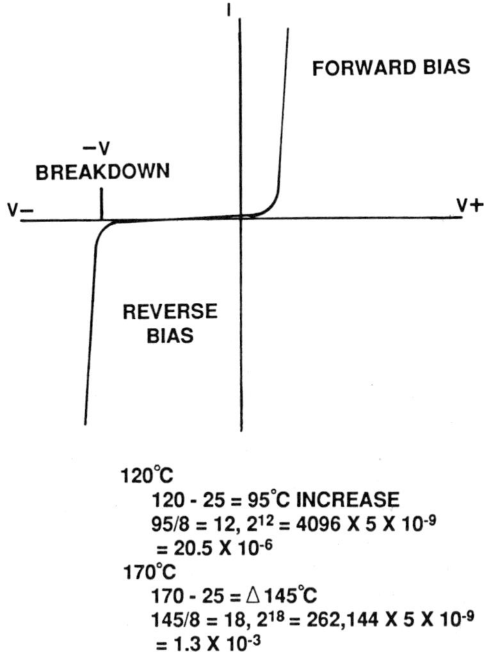



120°C
   120 - 25 = 95°C INCREASE
   95/8 = 12, $2^{12}$ = 4096 X 5 X $10^{-9}$
    = 20.5 X $10^{-6}$
170°C
   170 - 25 = $\triangle$ 145°C
   145/8 = 18, $2^{18}$ = 262,144 X 5 X $10^{-9}$
    = 1.3 X $10^{-3}$

FIGURE 5.5  PN junction characteristics.

## 5.5  TEMPERATURE-COMPENSATION TECHNIQUES

Temperature-induced errors of piezoresistive sensors can be separated into two categories: temeprature effects on offset (change
in the DC output voltage with no pressure applied) and temperature
effects on span (sensitivity to applied pressure). Although the
user is interested only in the combined effect, they must be treated
separately when temperature-compensating the sensor. There are
three techniques used to temperature-compensate piezoresistive
sensors: passive compensation at the sensor, active compensation
in the signal-conditioning electronics, and software compensation
when a microprocessor is used in the system. We will now discuss
the more common techniques used to compensate for temperature
errors.

### 5.5.1  Span Temperature Compensation

Span is defined as the gage factor (pressure sensitivity usually
given in mV/V/psi—millivolts of output change per volt of bridge

excitation voltage per psi of applied pressure) multiplied by bridge voltage and applied pressure:

$$S = GF \times V_b \times psi$$

where   S = span

GF = gage factor

$V_b$ = bridge voltage

psi = applied pressure

From the equation, if the bridge voltage is changed (gage factor and pressure remain constant), the span will change proportionally. The span is ratiometric to the bridge voltage. Also, if the gage factor changes, the bridge voltage can be adjusted to maintain a constant span. All the span temperature compensations use this technique. The gage factor of a piezoresistive sensor will have a slightly nonlinear-negative temperature coefficient (see Figure 5.6). Increasing the bridge voltage at the same rate as a function of temperature results in a constant span over temperature.

The two most popular compensation techniques differ in the way temperature is measured. One technique is to apply constant current excitation and use the bridge resistance to control the bridge voltage. The other technique is to use a temperature-sensitive element (thermistor, diode, transistor, or temperature-sensitive current source) that is not on the sensor chip to measure temperature.

*Constant Current Source*: A constant current source is a circuit
   with high output impedance that maintains a set current to
   the load independent of the load impedance or voltage developed
   by the load.
*Constant Voltage Source*: A constant voltage source is a circuit
   with minimum output impedance that maintains a constant output
   voltage independent of load impedance or current supplied to
   the load.


## 5.5.2  Constant Current Excitation

The advantage of sensing temperature with the sensor chip itself is that it senses not only the ambient temperature, but also the temperature of the working medium. In some applications, a hot or cold liquid may be in contact with the die, causing large temperature gradients from ambient temperature to sensor temperature. If

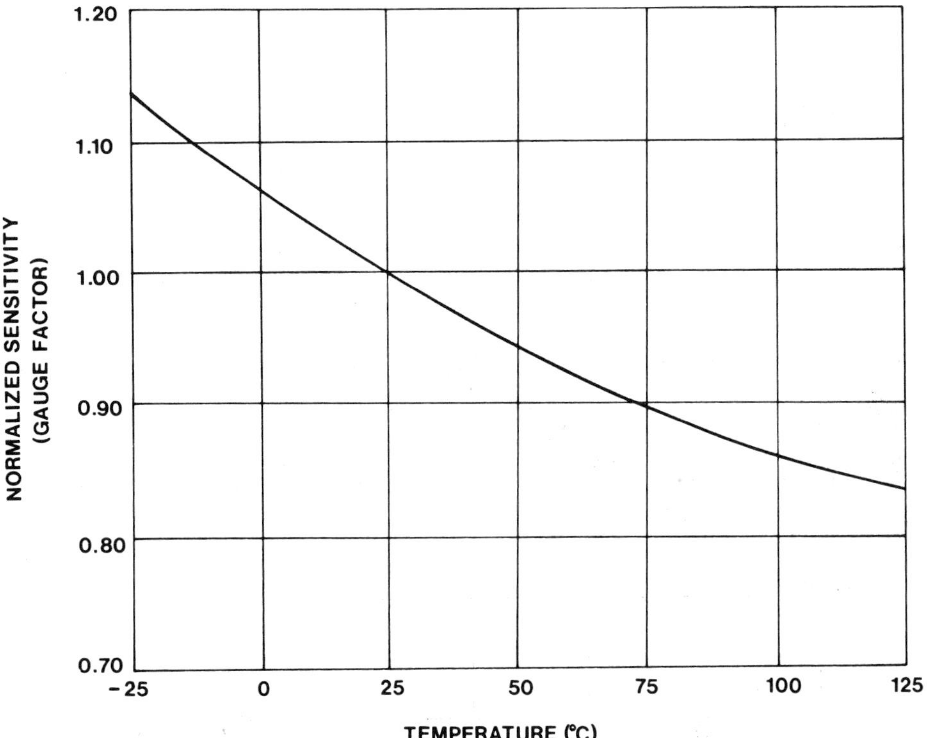

FIGURE 5.6  Sensitivity vs. temperature.

the temperature coefficient (TC) of the bridge resistors is equal
to or greater than the temperature coefficient-gage factor (TC-GF),
the span can be temperature-compensated by driving with a constant
current and forcing the bridge impedance TC to equal the TC-GF.
To use this technique, the temperature-coefficient resistor (TCR)
must be equal to or greater than the TC-GF. This is not the case
for all sensors on the market, so constant current excitation is
not always an option. Figure 5.7 shows the relationship between
gage-factor temperature coefficient and bridge resistor coefficient.

A current-excitation circuit can be designed in a number of
ways. IC current sources are available for some fixed currents,
or a current source can be designed from an operational amplifier
and transistor or field-effect transistor (FET) (see Figure 5.8).
Putting the sensor into the feedback of the operational amplifier
is the most popular method. Figure 5.8a is the circuit used with
sensors. $R_2$ sets the bias current for reference diode $Z_1$; $Z_1$ should

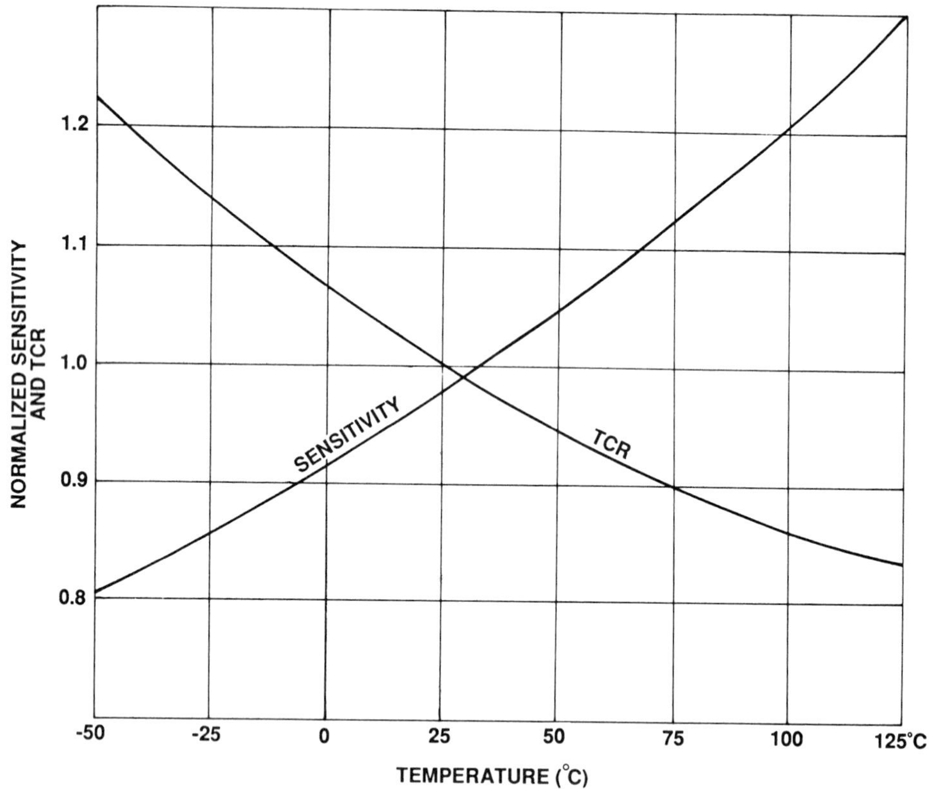


FIGURE 5.7   TC sensitivity vs. TCR.

be a stable, low-TC reference diode. Band-gap reference diodes
are available that operate over large current ranges, have less
than 1-ohm input impedance, and are very stable over temperature.
Resistor $R_1$ sets the current through the bridge: $I_B = V_r/R_1$.
$R_1$ should be a stable, low-TC resistor. Considerations when designing this circuit include voltage compliance (maximum voltage swing
of the amplifier) and bridge-impedance changes with temperature.

Figure 5.8b shows another circuit that uses an operational amplifier to generate a high-output-impedance current source. Here
again, $R_2$ and $Z_1$ generate a stable reference voltage for the circuit
and $R_1$ sets the output current. The difference in this circuit is
that the amplifier output is buffered by a PNP transistor $Q_1$, and
the load is connected to the high-output impedance collector of
the transistor. An FET, which has a high input impedance, can
also be used as a buffer.

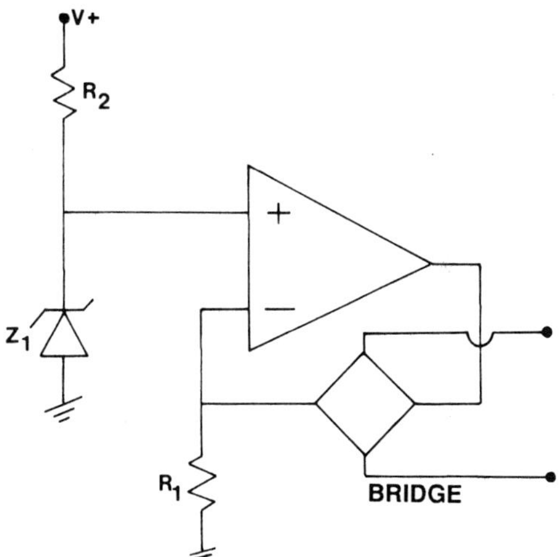



**(A) SENSOR IN FEEDBACK OF OPERATIONAL AMPLIFIER**

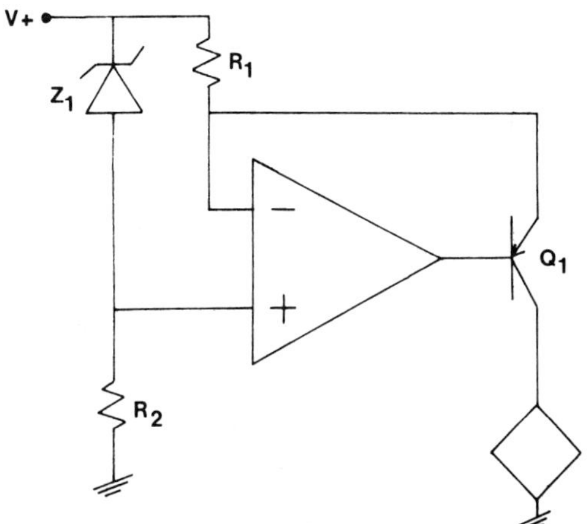



**(B) HIGH OUTPUT IMPEDANCE CURRENT SOURCE**

FIGURE 5.8　Constant current sources.

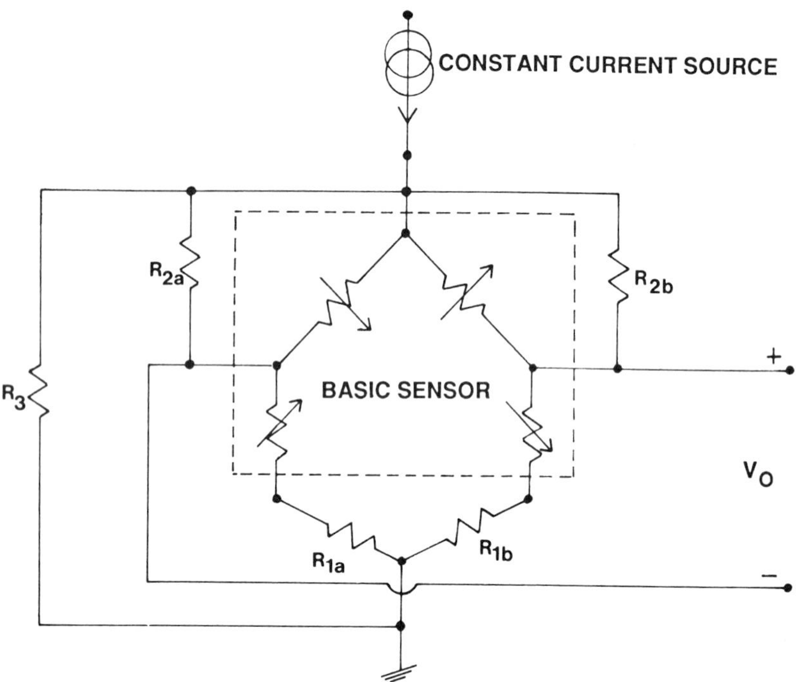


FIGURE 5.9   Passive sensor compensation.

For a constant-current application, the bridge voltage is the product of the bridge resistance and the applied current. The basic sensor will be overcompensated (have a positive span TC), and the TCR of the bridge will have to be decreased by shunting the bridge with a low-TC resistor, reducing the span compensation. Figure 5.9 shows the passive compensation technique used with constant current. The compensated device will consist of the basic sensor and three external resistors. $R_1$ is chosen to adjust the offset voltage. If the offset is positive, $R_{1a}$ is used and $R_{1b}$ is shorted. If the offset is negative, $R_{1b}$ is used and $R_{1a}$ is shorted. $R_2$ is the offset temperature compensation resistor, which decreases the temperature coefficient of the appropriate bridge resistor and changes the sensor offset as a function of temperature. Here again, only one resistor is used and the other is left open. $R_3$ is the span-compensating resistor shunting the entire bridge. Since a nonlinear function is used to compensate for another nonlinear function, the shape of the span/temperature curve will be a function of how closely the two curves match.

## 5.5.3  Constant Voltage Excitation

The same basic technique is used to compensate voltage-driven
sensors. The voltage supplied to the sensor is changed as a function
of temperature and gage-factor temperature characteristics. If
TCR > TC-GF, a resistor can be placed in series with the sensor,
reducing the equivalent TCR of the bridge. If TCR < TC-GF, the
equivalent of a temperature-controlled voltage source needs to be
designed. Passive components such as thermisters (temperature-
sensitive resistors), series diode or transistor networks, or a
temperature-sensitive voltage can be summed to a reference voltage
using an operational amplifier. The main difference is that the temperature is sensed by some component other than the bridge impedance. If a true temperature-sensitive voltage source is designed,
any sensor can be used, since the voltage source will compensate
for any changes in bridge impedance and change the bridge voltage
as a function of the temperature variations.

## 5.5.4  Thermistor Compensation

If the TCR of a sensor is less than the TC-GF (TCR = 600 to 1800
ppm, TC-GF = -2000 to -2300 ppm), a temperature-sensitive device
must be added to the circuit to compensate span over temperature.
   Assume a TCR of 1200 ppm and a TC-GF of 2200 ppm. If the
sensor is driven by a constant current source,

$$TCS + TCR = 0$$
$$-2200 + 1200 = 0$$
$$-1000 = 0$$

The span is undercompensated by -1000 ppm, or -0.1 percent per °C.
   If the sensor is powered by a power supply (voltage source),
a thermistor network with a negative temperature coefficient can
be placed in series with the sensor to create a voltage divider with
a positive temeprature coefficient at the top of the bridge. This
is not a true voltage source at the top of the bridge—the bridge
impedance and TCR of the bridge affect the voltage across the
bridge. Fixed thermistors are available with different temperature
characteristics, or thermistor inks are available for screening thick-
film resistors. Thermistors are nonlinear over temperature (see
Figure 5.10), but the temperature characteristics can be modified
by adding series and shunt resistors:

$$R_T = \text{thermistor} \qquad R_T = R_0\, e^{\beta(1/t_0 - 1/t)}$$

where $R_O = R_T$ at the reference temperature $t_0$ and $\beta$ is a constant. The overall resistance is

$$R_N = (R_T + R_1)R_2/R_T + R_1 + R_2$$

The overall impedance change as a function of temperature T is

$$\Delta R = (\beta/T^2)\,(R_2)^2\,R_T/(R_1 + R_2 + R_T)^2$$

If the bridge TCR and TC-GF are known, the thermistor TCR necessary to compensate the bridge can be calculated. These sensors are generally assembled, placed in a temperature chamber, and measured over the temperature range of interest. Offset, offset TC, span, and span TC data are measured and stored. These data are then used to adjust the compensation circuits to obtain the desired characteristics. Thick-film networks containing resistors and thermistors are common in high-volume applications; these networks can be laser-trimmed using a computer-controlled laser. The stored data are read by the computer, which makes the necessary

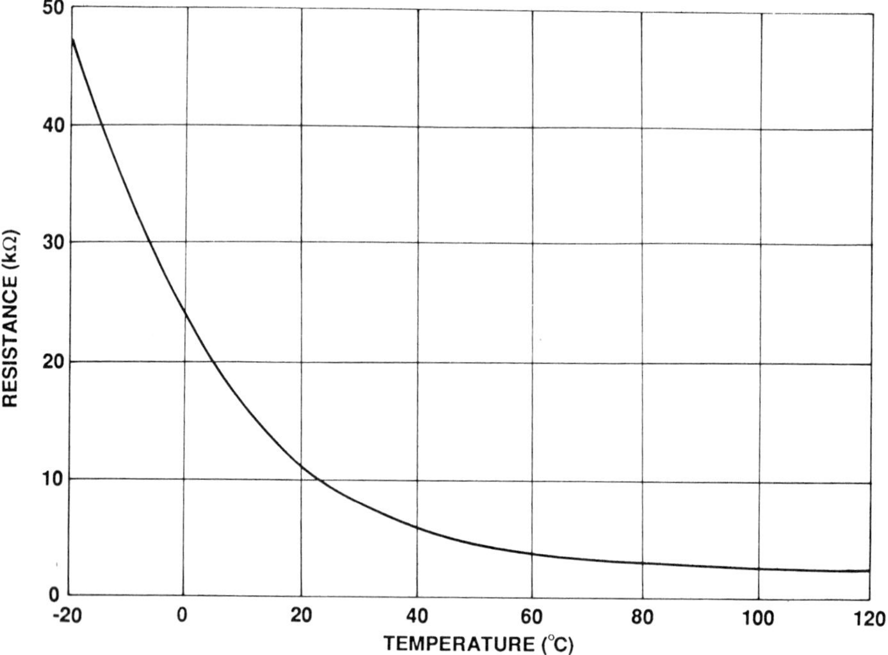


FIGURE 5.10   Thermistor characteristics.

calculations and controls the laser-trimming operation. Since a non-linear function is used to compensate another nonlinear function, the shape of the span-vs.-temperature curve will not be a straight line.

This span-compensation technique has two disadvantages. First, a relatively large voltage is dropped across the thermistor network, decreasing the voltage available to the bridge. Second, the compensation circuit is on a ceramic that is in the same package but not thermally coupled to the sensing die. In fact, if a plastic or glass-filled nylon package is used, a high thermal resistance is placed between the ceramic and sensor die.

### 5.5.5  Diode String Compensation

A low-cost, moderately accurate span-compensation technique is to use general-purpose silicon diodes in series with the sensor. Silicon diodes have a very predictable temperature coefficient of -2500 ppm/°C (-0.25 percent/°C) when forward-biased (see Figure 5.11). A number of diodes can be used to generate a bridge voltage with a positive temperature coefficient equal to TC-GF:

$$V_B = V_S - n\phi$$

$$\Delta V_B = -n(\Delta\phi)/(V_S/\phi - n)$$

where $V_B$ = bridge voltage

$\quad\quad$ n = number of diodes

$\quad\quad$ $V_S$ = supply voltage

$\quad\quad$ $\phi$ = forward-biased diode (0.7 V)

$\quad\quad$ $\Delta\phi$ = -2500 ppm/°C diode TC

*Example*: What will the span TC be for a sensor with -2300 ppm (TC-GF, 6-V supply, and four diodes?

$$V_B = 6 - (4)(0.7)$$

$$V_B = 3.2 \text{ V}$$

$$\Delta V_B = -4(2.5 \times 10^{-3})/6/0.7 - 4$$

$$\Delta V_B = 2188 \text{ ppm/°C}$$

The span-TC error will be 2300 - 2188, or 112 ppm, which is acceptable.

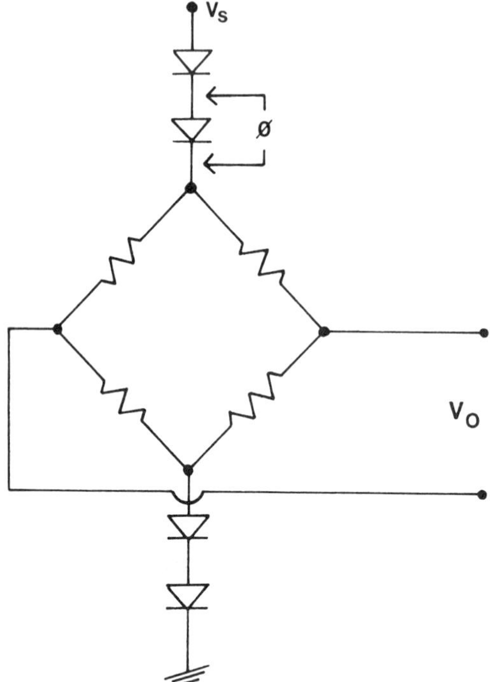


FIGURE 5.11   Diode span compensation.


    The disadvantages of this technique are that half the power-
supply voltage is dropped across the compensation network, whole
diodes have to be added or subtracted from the network, and it
is difficult to thermally couple the diodes to the sensing chip.


## 5.5.6   Transistor Compensation

An NPN transistor and two resistors can be placed in series with
the bridge. The transistor is used to amplify the voltage developed
across the forward-biased base-to-emitter diode (see Figure 5.12).
The ratio of the two resistors sets the nubmer of diodes ($\alpha$):

$$\alpha = 1 + R_1/R_2$$

$$V_B = V_S - \alpha\phi$$

$$\Delta V_B = -\Delta\phi \ (\alpha/(V_S/\phi) - \alpha$$

where $\alpha$ = number of diodes

$V_B$ = bridge voltage

$\phi$ = forward-biased diode drop

This circuit will compensate for the ±10 percent changes in bridge impedance; however, if a different bridge is to be used, the resistor values should be checked. If the circuit is set up for a 500-$\Omega$ bridge, a 5000-$\Omega$ bridge may not draw enough current to forward-bias the transistor.

The advantages of this circuit over the diode string is that the resistors can be adjusted with potentiometers or laser-trimmed to match the sensor. The disadvantages are that both the diode and the transistor technique will work for a specific power-supply voltage, but if the supply should change, the compensation has to be recalculated. This device is also difficult to thermally couple to the sensor.

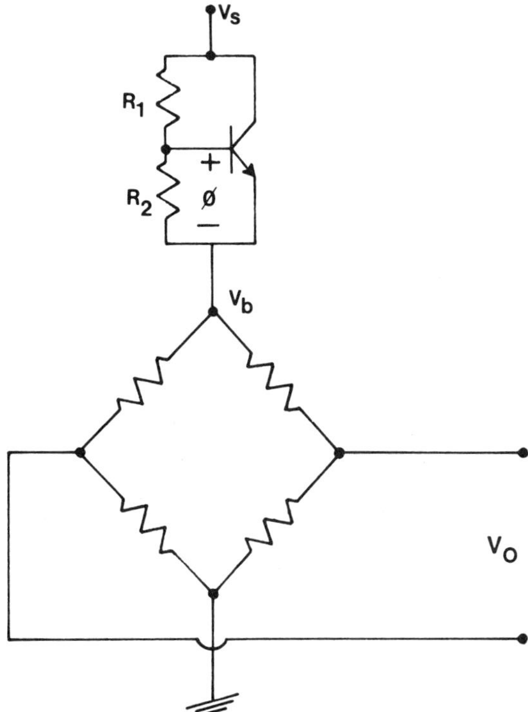


FIGURE 5.12  Transistor span compensation.

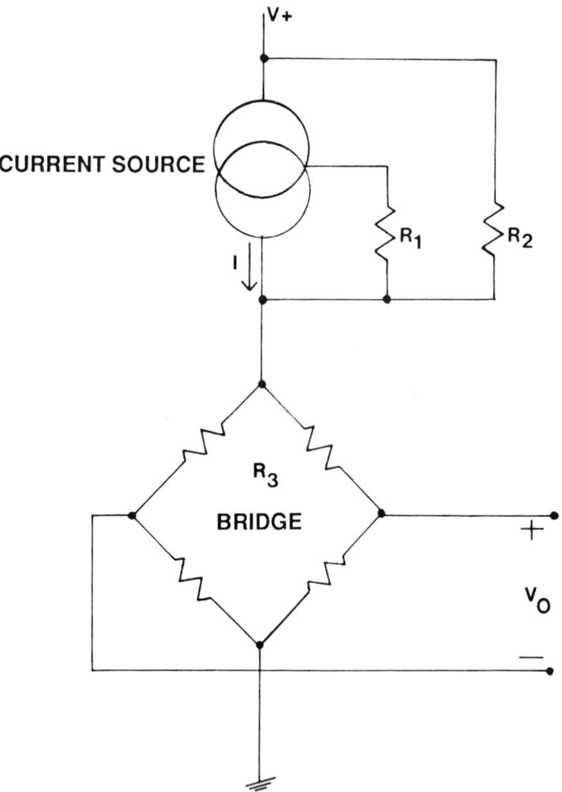

**FIGURE 5.13**  Temperature-controlled current source.

### 5.5.7  Temperature-Sensitive Current Source

Temperature-sensitive current sources are available that can be
used to compensate span TC. These current sources have a large
temperature coefficient, on the order of 3360 ppm, which is too
large to compensate a sensor. The current-source TC can be decreased by shunting with a fixed resistor (see Figure 5.13). The
circuit shown in Figure 5.13 is a three-terminal monolithic current
source. $R_1$ sets the current and $R_2$ modifies the temperature coefficient of the circuit:

$$V_B = \alpha(V_S + I\,R_2)$$

$$\alpha = R_B/R_2 + R_B$$

$$I = 67.7 \text{ mV}/R_1$$

$$\Delta V_B = \Delta R_B (1 - \alpha) + \Delta I [1 - \alpha (V_S/V_B)]$$

where  I = current-source current

$\Delta I$ = 3360 ppm

$\Delta R_B$ = bridge TCR

The advantage of this circuit is that the voltage across the current source will be about 1 V, allowing more voltage to be developed across the sensor. Like the diode and transistor compensation, the current source should be thermally coupled to the sensor.


### 5.5.8  Offset-Temperature Compensation

The offset voltage is the output voltage of an unstrained bridge— 0 psig for gage and differential sensors, or 0 psia for absolute sensors. Offset-temperature coefficients confuse some people. A Wheatstone bridge will not have an offset error if the four resistors are matched and track over temperature. This is true if we consider just an electrical circuit; a pressure sensor, however, is an electro-mechanical device and reacts to mechanical inputs as well. The bridge will react to the normal electrical parameters such as resistor tolerances and thermal tracking, as well as to mechanical stresses such as strains induced into the chip during processing and stresses transmitted through the mount from the package. The offset-temperature coefficient will be the sum of all these inputs.

*Electrical*

*Resistor Mismatch*:  If the bridge has an initial offset due to resistor mismatch (not balanced), the bridge will have a thermal error equal to $V_{offset} \times 2300$ ppm (span compensation)

*Resistor Tracking*: How closely the resistors track over temperature.

*Self-Heating*:  The area occupied by the resistors is small compared to the area of the sensing die, and hot spots are generated. This effect is most noticeable at turn-on and if the thermal resistance of the medium changes. If a sensor is measuring 1 atmosphere (14.7 psi) and the pressure is changed to 150 psi, the air is compressed and the sensor cools slightly. If the pressure is returned to 1 atmosphere, an offset error is present. This offset will decay to zero as the chip thermally stabilizes; this is referred to as *warm-up error*. Battery-operated systems that apply power, take a measurement, and turn the power off are affected by warm-up.

*Temperature Gradients*:   Due to the large temperature coefficient
   of the bridge resistors, a small thermal gradient across the
   chip can create a large offset error. The other source of error
   is the thermal gradient from the sensor chip to the conditioning
   electronics. If the temperature is changed rapidly, a large
   initial offset generally occurs and then decays as the components
   thermally stabilize. These errors cannot be compensated; they
   can only be minimized.

*Bridge Loading*:   If the signal-conditioning electronics have a low
   input impedance or the differential input impedance is not balanced, the sensor offset TC can be affected.

*Mechanical*

*Process-Induced*:   As the wafer is processed, strains are induced
   into the silicon by the high-temperature processing.

*Mismatches in Coefficient of Expansion*:   The sensor mount can
   generate offset-temperature errors if the thermal coefficient
   of expansion is significantly different from that of silicon. Also,
   strains generated by the package material can be coupled to
   the chip through the mount.

## 5.5.9  Passive Compensation

The large TCR of the bridge resistors is also used to compensate
for offset-temperature coefficients. If one of the four bridge resistors
is shunted with a low-TC resistor or a small resistor is placed in
series with a resistor, both offset and offset TC are affected. By
knowing the TCR of the sensor and measuring its offset TC, a
resistor value can be calculated to compensate for offset TC. Some
sensors are sold with a computer printout that indicates the value
of the three compensation resistors. Each sensor will have different
compensation resistors; if the sensor is changed for some reason,
the compensation resistors must be changed as well.

   If the sensor has a compensation ceramic, sets of TC resistors
are silk-screened on the ceramic—one for positive errors, the other
for negative. A thick-film resistor can only be increased in value
by laser-trimming. The compensation circuits are designed using
the ratio of two resistors, or one resistor is opened by laser-trimming
to be able to increase or decrease the amount of compensation.
These are stand-alone devices; the compensation ceramic is soldered
to the sensor and is part of the package.

   Another approach is to generate a temperature-sensitive voltage
using thermistors or some other temperature-sensitive device and
then to couple this voltage to the bridge through a large-value

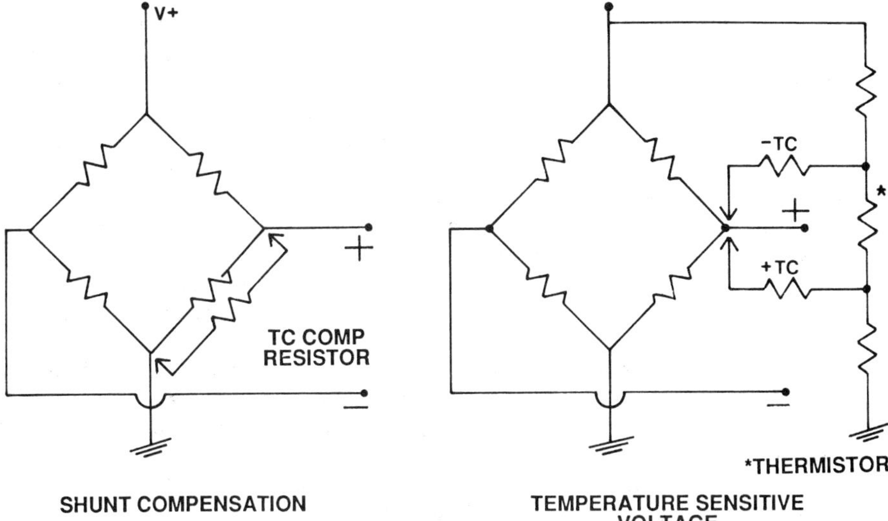



FIGURE 5.14   Offset TC compensation.

resistor (see Figure 5.14). Each sensor has to be tested over the
temperature range of interest to determine the amount and sign
of the compensation required.


## 5.5.10   Active Compensation

Active compensation uses operational amplifiers to fix the amount
of compensation and to sum the error signal with the sensor output.
This technique has the advantage that the entire circuit can be
offset-temperature-compensated by summing the correction signal
at the circuit output, not at the sensor signal-conditioning amplifier.
This technique will work with both current-driven sensors and
voltage-driven sensors. The most conventient temperature signal
is the bridge voltage. We know the bridge voltage will have a positive
temperature coefficient of 2100 to 2300 ppm to compensate for span
TC. This voltage can be amplified or attenuated and summed with
the sensor output to give minimum offset error due to temperature.


## 5.5.11   Auto-Zero Techniques

If a known reference pressure is measured by the sensor at some
step in the operation, an auto-zero circuit can be used to zero out

all offset errors. (Figure 5.15 shows the basic block diagram of an auto-zero circuit.) An auto-zero circuit can be used with a basic sensor or with a sensor that has been compensated, but does not meet the desired accuracy. The auto-zero circuit can be incorporated into the amplifier stage following the sensor, or into any amplifier in the circuit. As shown in Figure 5.15, the output voltage is compared to a reference voltage by a high-gain device (integrator or comparator, depending on the type of sample and hold used). This voltage is fed to the sample-and-hold stage, which is controlled by a logic signal to put it into the sample or hold mode. The output of the sample and hold is summed with the sensor output by an operational amplifier.

The key to the auto-zero circuit is the sample-and-hold circuit. This function can be done digitally or with an analog sample and hold. Many IC sample-and-hold amplifiers are available on the market; these generally charge a capacitor to store the voltage and a high-input impedance amplifier to buffer the capacitor from the rest of the circuit. If the sample-and-hold circuit is not required to hold a voltage for a long period of time (minutes or longer), the devices are very cost-effective. If the device is left in the hold mode for long periods of time, the charge on the capacitor will change due

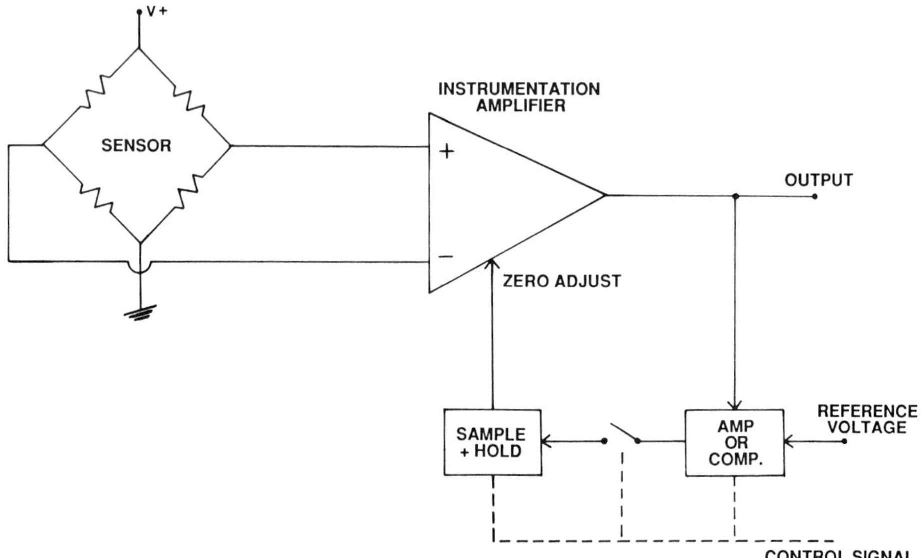


FIGURE 5.15   Basic auto-zero circuit.

to current leakages in the circuit. Another approach is to use a digital-to-analog (D/A) converter as a sample-and-hold circuit. The digital-to-analog converter is the equivalent of a digitally controlled power supply. The output voltage will not change as long as the digital input is unchanged. The accuracy of the system will determine the D/A converter used: an 8-bit has 1/256 resolution, a 10-bit has 1/1024 resolution, and a 12-bit has 1/4096 resolution. A D/A circuit would consist of a comparator to sense the output voltage and supply a logic signal if the output is greater or less than the desired reference voltage; an oscillator (clock); a binary counter that controls the D/A converter; and a D/A converter that generates the correction voltage necessary to give the desired output voltage (Figure 5.16). The sequence of events is:

1.  The reference pressure is applied to the sensor.
2.  A control signal resets the counter to zero, forcing the output to always be less than the reference voltage.
3.  The comparator output goes low, allowing clock pulses to be counted by the counter.
4.  This process continues until the output voltage equals the reference voltage and the comparator goes high, inhibiting the clock pulses and putting the circuit in the hold mode. The circuit stays in the hold mode until another reset pulse restarts the sequence.

A latch is used in the logic that controls the clock pulses to the counter to ensure that clock pulses are not counted every time the output is less than the reference voltage.

This is the basic circuit, and many variations can be built into the logic and control functions. Up-down counters could be used with a window comparator to make a tracking D/A converter that does not have to be reset every time. Also, if the reference voltage is the lowest voltage used, the circuit can be designed to zero any output less than the reference voltage (any output less than the reference is an error voltage). This circuit can also be used as a peak-and-hold circuit to record just the maximum pressure attained by the system.

## 5.5.12  Software Compensation

Software compensation can be classified into two groups: auto-reference and lookup table. The application will determine which, if any, software compensation should be used.

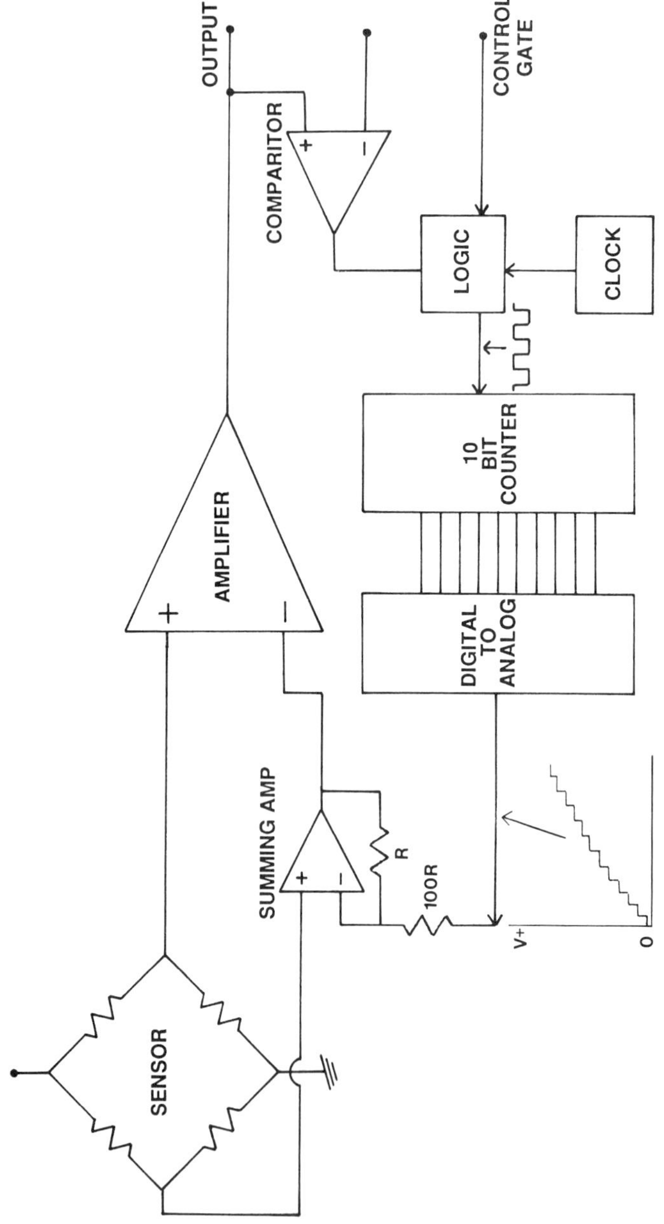



FIGURE 5.16  Digital auto-zero circuit.

*Auto-Reference*:   Auto-reference is similar to auto-zero, in that
   a reference pressure is applied to the sensor and the output
   voltage is measured and stored in memory. When different pressures are measured, the stored reference voltage is subtracted
   from the reading, giving the true span and compensating for
   any offset errors. This technique will not compensate for any
   sensitivity errors. Any system that is cyclic—such as blood-
   pressure or tire gages where a reference condition of zero
   pressure precedes the actual measurement—is ideal for this
   technique.

*Lookup Table*:   Lookup-table compensation requires more data inputs,
   such as temperature, but gives a more accurate reading. To
   use this technique, the sensor is measured over the temperature
   and pressure range of interest. These data are then put into
   a nonvolatile memory. At each reading, the temperature and
   sensor output is measured and read by the microprocessor;
   the microprocessor then reads the proper output for that temperature and sensor reading. This technique is more accurate in
   that it compensates for both offset and span errors. To use
   this technique, the sensor must have good repeatability—it
   has to have the output voltage at a given temperature and
   pressure time after time. The disadvantage to this approach
   is that it requires large amounts of memory to get adequate
   resolution. This type of compensation is ideal for systems that
   are nonlinear, such as flow meters or altimeters. The output
   voltage can be converted to flow rate or altitude and displayed
   in the proper units.

## 5.6  SIMULATION OF DIAPHRAGM-TYPE IC PRESSURE SENSORS (See Reimann and Fathi, 1986)

Monolithic silicon pressure-sensor chips with piezoresistive sensing
elements on a michromachined diaphragm are commercially available
from many manufacturers. New developments are in progress to
meet the requirements of small (catheter-tip) sensor, low-pressure-
range sensors (high sensitivity, 0.1-psi pressure range), or the
integration of compensation circuitry. Computer-aided design is
replacing the time-consuming evaluation of new chips through the
processing of prototype sensors. Sensor simulation allows the designer
to consider different layouts, use the results in an interactive way
to optimize the design, and test the stability of the design with
regard to fabrication tolerances.

The calculation of the sensitivity of a piezoresistive pressure
sensor is performed in three steps:

1.  *Stress Calculations*:   Given the geometry of the structure,
    the elastic properties of the material, and the applied pressure
    difference, the stress distribution in the diaphragm is calculated.
2.  *Piesoresistive Transduction Mechanism*:   The changes in the
    resistivity as a function of the stress components are measured.
3.  *Sensor Response*:   The resistive changes are translated into
    changes in the differential output voltage of the sensing element.

Pressure sensors manufactured by anisotropic etching of (100)-
silicon are considered. The etch stop on the (111)-planes leads
to a rectangular diaphragm constrained by a rigid frame. Figure 5.17
shows the geometry and introduces the coordinate system. The diaphragm, with dimensions L × W × H, is determined by the wafer
thickness T and the chip-cavity mask $L_m$ × $W_m$: W = $W_m$ − SQR(2) ×
(T − H). L/W is the aspect ratio. Normalized coordinates x' = 2X/L

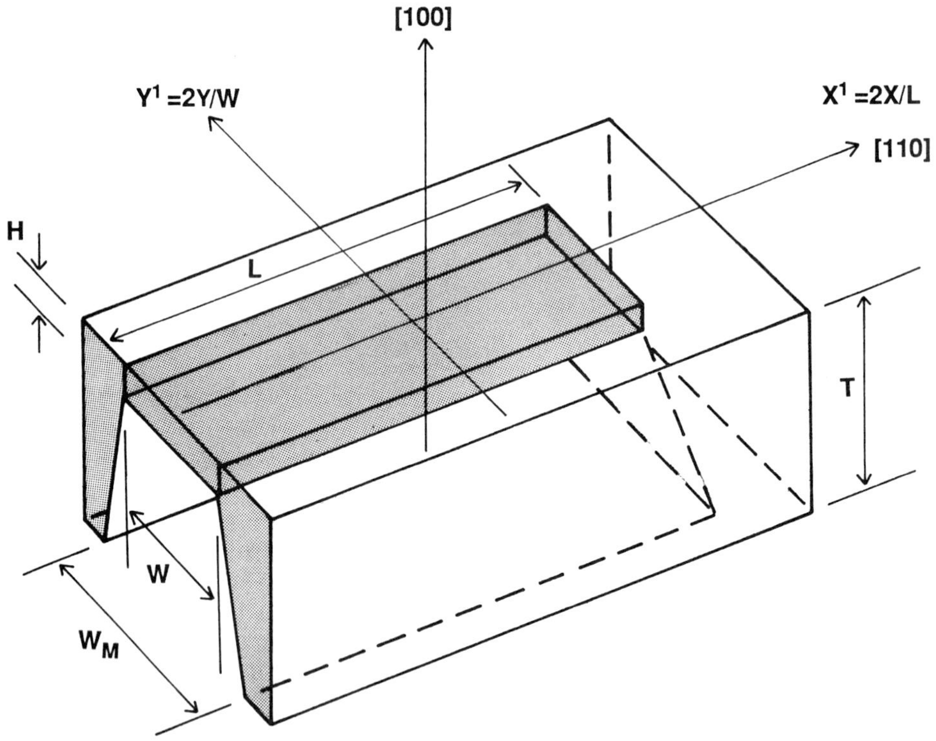


FIGURE 5.17   Pressure-sensor chip micromachined in 100-silicon.

and $Y' = Y/W$ define a point on the diaphragm. The basic assumptions of the model are:

1. *Diaphragm*: Thin plate of dimensions $L \times W \times H$, edges ideally clamped, boundary nonlinearity neglected.
2. *Small Displacements*: Only lateral forces considered, geometric nonlinearities neglected.
3. *Anisotropic Material*: Material constants of silicon.
4. *Etch Stop on (111)-Plane*: For a wafer thickness T, mask length M, and diaphragm thickness H, the length of the diaphragm is

$$L = M - 2(T - H) \tag{5.1}$$

Given the deflection of the diaphragm, kinematics determines strain and stress components. Unfortunately, an analytical solution exists for circular plates only. For rectangular plates, variational methods using a series expansion for the deflection (see Timoshenko and Woinowsky-Kreiger, 1959) or numerical integration using finite differences (see Lee and Wise, 1982) must be used.

### 5.6.1 Stress Analysis

The starting point for the analysis is an expression for the deflection $W(x', Y')$ of the diaphragm. This function is written as a super-position of base function $\phi$, which has proper symmetry and boundary conditions. The coefficients of the series expansion are determined by the minimum of the potential energy (Ritz-Galerkin method). Timoshenko (see Timoshenko and Woinowsky-Freiger, 1959) has worked out the problem with four base functions:

$$\phi_1 = ((x')^2 - 1)^2 \times ((Y')^2 - 1)^2; \quad \phi_2 = (Y')^2 \times \phi_1$$
$$\phi_3 = (x')^2 \times \phi_1; \quad \phi_4 = (x')^2 \times (x')^2 \times (Y')^2 \times \phi_1 \tag{5.2}$$

and has presented the solution for the square plate with isotropic elastic constants.

In this work, the same series expansion is used for the rectangular plate. The expression for the deflection is

$$W = (x', Y') = W(0,0) \times F(x', Y')$$
$$F(x', Y') = ((Y')^2 - 1)^2 ((x')^2 - 1)^2 \{1 + C_1 (Y')^2$$
$$+ C_2 (x')^2 + C_3 (Y')^2 (x')^2\} \tag{5.3}$$

where $W(0,0) = C_0 \times W^4 \times P/D$ is the deflection at the center of the plate, F is a normalized function for the deflection, P is the

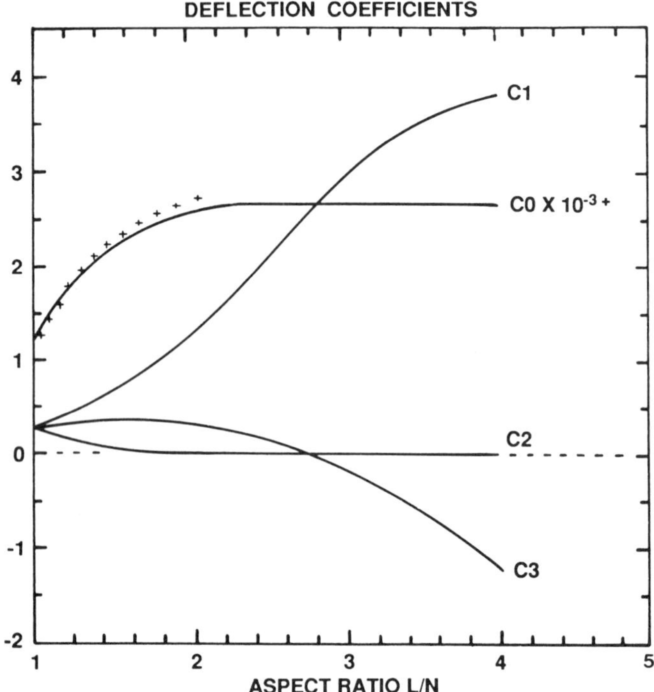



FIGURE 5.18   Deflection coefficients for a rectangular plate.

pressure difference across the plate, and $D = E' \times H^3/(12(1 - (v')^2))$
is the flexural rigidity. The geometry is defined in Figure 5.17.
The four coefficients $C_0$ ... $C_3$ are determined by a least-squares
fit to the results of an independent finite-element analysis (FEA).
Abaqus (see Hibbit, Karlsson, and Sorenson, 1982) was used as
a finite-element code. One-fourth of the plate was modeled with
a net of $4 \times 4$ and $10 \times 10$ second-order shell elements. This analysis
was performed for a series of plates with aspect ratios between 1
and 4. For each aspect ratio, the deflection was fitted with Equation
5.3 (see Hibbit, Karlsson, and Sorenson, 1982), and the resulting
deflection coefficients were again fitted with cubic splines to determine the general dependence on the aspect ratio. Figure 5.18 shows
the results, and the following table gives selected values for the
square and one rectangular plate.

| | L/W | $C_0$ | $C_1$ | $C_2$ | $C_3$ |
|---|---|---|---|---|---|
| Timoshenko | 1 | $1.264E^{-3}$ | 0.264 | 0.2645 | 0.309 |
| FEA (Si) | 1 | $1.38\ E^{-3}$ | 0.219 | 0.219 | 0.320 |
| FEA (Si) | 2 | $2.64\ E^{-3}$ | 0.444 | 0.0223 | 0.338 |

Kinematics determines the strain components $\varepsilon_K$ as the second derivatives of the deflection and the stress components $\delta_K$ as linear combinations of the strain components.

The effective elastic moduli for the coordinate system of Figure 5.17 are $E_{xx'} = 1.70$, $E_{xy'} = 0.107$, Young's modulus $E' = 1.69$ (all in units of $10^{12}$ dyn/cm$^2$), and Poisson ratio $V' = 0.063$ (Wortman and Evans, 1965). The results are written as the product of an amplitude factor and two normalized functions $\phi_{xx}$ and $F_{yy}$ describing the strain distribution across the plate.

$$\delta_x(x',Y') = A\ (F_{xx} + V \times F_{yy})$$
$$\delta_y(x',Y') = A\ (V \times F_{xx} + F_{yy}) \tag{5.4}$$

With $A = 48 \times C_0 \times f(z') \times P \times (W/H)^2$,

$$F_{xx}(x',Y') = a^{-2}((Y')^2 - 1)^2\{[C_2 - 2] + (Y')^2 \times [C_3 - 2C_1]$$
$$+ 6(x')^2([1 - 2C_2] + (Y')^2[C_1 - 2C_3])$$
$$+ 15(x')^4[C_2 + (Y')^2 \times C_3]\}$$
$$F_{yy}(X',Y') = ((x')^2 - 1)^2\{[C_1 - 2] + (x')^2[C_3 - 2C_1] \tag{5.5}$$
$$+ 6(Y')^2([1 - 2C_1] + (x')^2[C_2 - 2C_3]$$
$$+ 15(Y')^4[C_1 + (x')^2 \times C_3]$$

$f(z') = 1 - (2 \times z'/H)$ describes the linear decrease from the surface ($z' = 0$) to the neutral plane, and $a = L/W$ is the aspect ratio of the plane. The shear stress is needed for the coordinate system with (100)-axis. This involves a rotation of 45 degrees and leads to the result $\delta_S = (\delta_x - \delta_y)/2$. Figures 5.19 and 5.20 illustrate the normalized stress distributions for a square and a rectangular plate with aspect ratio 2.

### 5.6.2  Piezoresistance and Sensor Response

The piezoresistive tensor ($\pi$) relates the stress components to anisotropic changes of the resistivity. For the planar problem of the thin

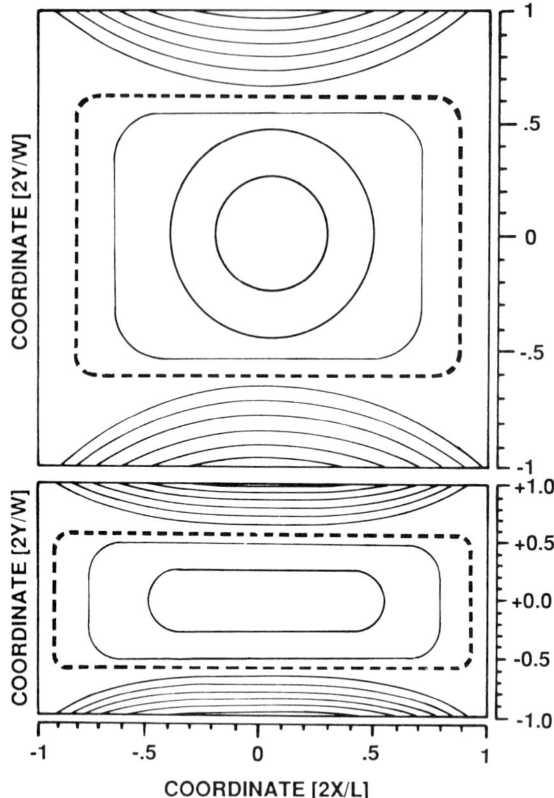

FIGURE 5.19   Contour plot for the stress component $\delta_y$ for square
and rectangular plate.


diaphragm, there are three lateral stress components and three
basic *sheet resistors*: $R_x$, a two-point resistor with current in the
X-direction; $R_y$, a two-point resistor aligned to the Y-direction;
and $R_{xy}$, a four-point resistor with current in the X-direction and
an electrical-field response in the Y-direction. r is used for the
stress-induced relative change in resistance. For the orientation
of practical importance, the piezoresistive tensor has the form

$$
\begin{array}{ccc}
\pi_e & \pi_t & 0 \\
\pi_t & \pi_e & 0 \\
0 & 0 & \pi_5
\end{array}
$$

This determines the two piezoresistors implemented in the simulation program.

1. The two-point resistor aligned parallel to one of the edges of the diaphragm:

$$r_x = \pi_e \delta_x + \pi_t \delta_4 \qquad\qquad r_y = \pi_e \delta_4 + \pi_t \delta_x$$

$$\pi_e = \tfrac{1}{2}(\pi_{44} + \pi_{11} + \pi_{12}) \qquad \pi_t = \tfrac{1}{2}(-\pi_{44} + \pi_{11} + \pi_{12}) \tag{5.6}$$

2. The four-point resistor aligned at an angle of 45 degrees toward the edge of the diaphragm:

$$r_{xy} = \pi_s \delta_s$$

$$\pi_x = \pi_{44}, \qquad\qquad \delta_s = \tfrac{1}{2}(\delta_x - \delta_y) \tag{5.7}$$

The value of the piezoresistive coefficients depends on the peak concentration of the resistor doping profile. The following values are consistent with published data (see Research Triangle Institute, 1964) for P-type silicon (units $10^{-12}$ $CM^2/dyn$):

$$\pi_{44} = 96.5\ (1 - 0.202\ \log\ (N/1E18))$$

$$\pi_{11} = 2.46\ (1 - 0.226\ \log\ (N/1E18))$$

$$\pi_{12} = -0.46\ (1 - 0.268\ \log\ (N/1E18))$$

In the simulation program, meander-like resistors can be laid out on the diaphragm. The geometry is broken down into squares of the line width, and the stress components are averaged over the number of squares.

As the last step, individual resistors are combined to a sensing element, and the output voltage is calculated. Two structures offer a differential output voltage ($V_{out}$), ratiometric to the supply voltage ($V_{in}$).

1. Wheatstone bridge of four two-point resistors:

$$V_{out}/V_{in} \simeq (r1 - r2 + r3 - r4)/4 \tag{5.8}$$

2. Shear element = one four-point resistor:

$$V_{out}/V_{in} = K_O \times r_{xy} \tag{5.9}$$

The constant $K_O$ depends on the geometry of the element. For a rectangular element with dimensions $L \times W$ and a homogeneous current

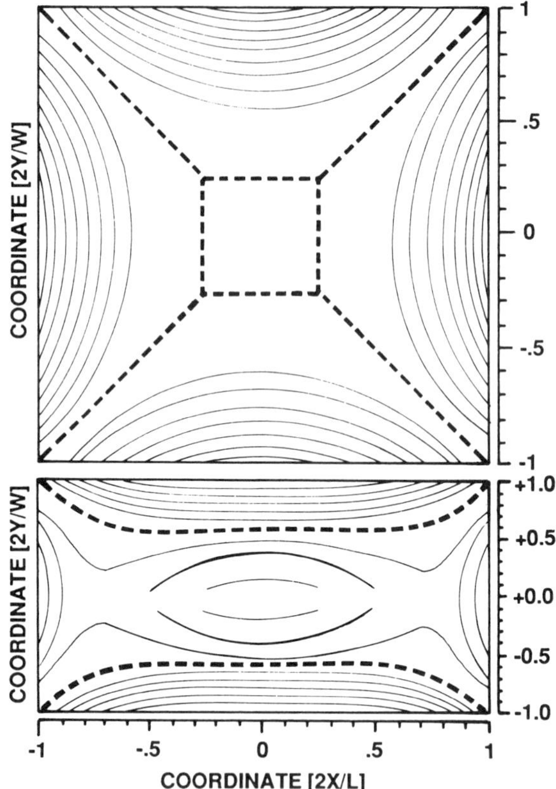

FIGURE 5.20   Contour plot for shear stress $\delta_S$ for square and
rectangular plate.

distribution, $K_O = W/L$. The transfer function $V_{out}/V_{in} = S \times P$
defines the sensitivity S of the device.

### 5.6.4   Simulation Results

For a given resistor layout, wafer thickness, and cavity mask,
the control of the diaphragm thickness leads to a series of chips
with different pressure ranges. The final result of the sensor simulation is a calculated curve for the sensitivity as a function of the
diaphragm thickness. This calculation takes the geometry of the
micromachined structure, looks up the deflection coefficients for
the given aspect ratio, calculates the stress distribution, averages

over the resistor layout, looks up the piezoresistive coefficients, and calculates the sensitivity. At this point, the response can be fitted to measured data points by adjusting a scale factor. Notice that there are no other adjustable parameters in the model.

Figure 5.21 shows a comparison of calculated and measured data. The data points are averages of a series of 10 devices, positioned at the specified diaphragm thickness. The example is for a Wheatstone bridge on a square diaphragm. The insert shows the resistor layout. The same kind of comparison was performed for a shear element on a square diaphragm and a Wheatstone bridge on a rectangular diaphragm. In each case, simulation and measurement agree within 10 percent (scale factor between 0.9 and 1.1).

The Timoshenko formula has been used by many groups for stress calculations of the square diaphragm. Its pseudoanalytical approach offers a fast and convenient solution for a desktop computer. Combining it with the finite-element method makes the approach accessible to anisotropic materials; rectangular diaphragms; and, with proper modifications, nonlinear corrections due to large deflections.

Typical agreement between the original finite-element analysis and the values calculated with the fitted series expansion is within 1 percent for the stress along the Y-axis, but reaches 10 percent for values along the length of the rectangular diaphragm. Minor strain components can be off by a factor of two. Going from $4 \times 4$ to $10 \times 10$ elements in the FEA changes the coefficients by a few

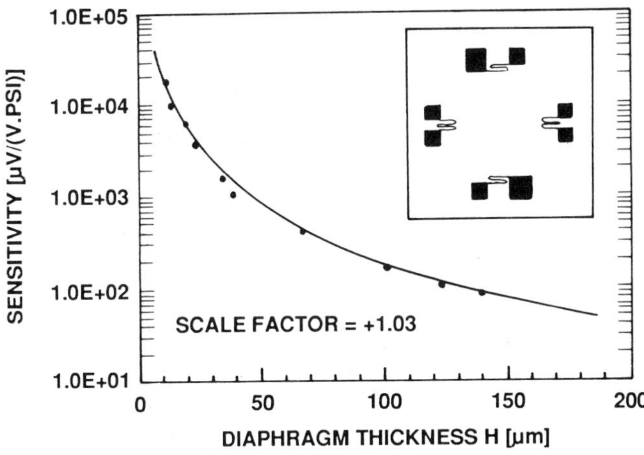


FIGURE 5.21 Comparison of measurement and simulation for a series of pressure-sensor chips.

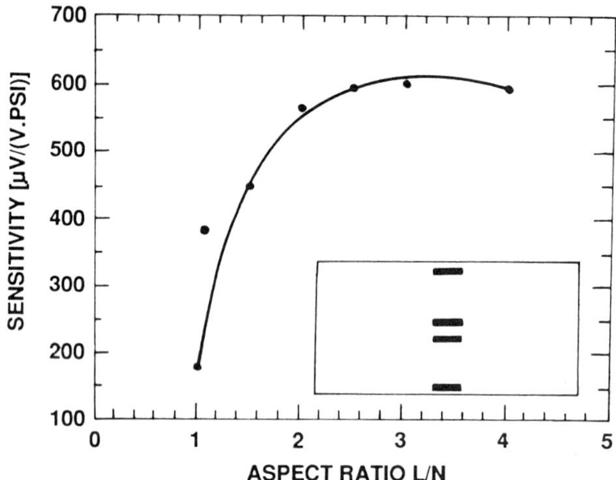

FIGURE 5.22   Sensitivity vs. aspect ratio for a given resistor layout
and constant diaphragm width.


percent. The agreement between measured and calculated sensitivities
is typically within 10 percent. The major uncertainty is the exact
thickness of the diaphragm.

The inherent strength of a simulation program is its interactive
use in the design and data-analysis process. Only a few examples
can be presented here:

1.   Figure 5.22 shows the transition from a square to a rectangular
     diaphragm for a given resistor layout (see insert) and constant
     diaphragm width. The sensitivity increases due to the contribution
     of the center resistors; it levels off when the stress distribution
     approaches that of a bridgelike structure across the width of
     the diaphragm.
2.   Manufacturing variations due to changes in wafer thickness,
     finite etch rate in the (111)-direction, mask registration, etc.,
     can be simulated. A point-like shear element near the edge
     of the diaphragm is most sensitive, while a design like the
     one in Figure 5.22 rejects many of these variations.
3.   The nonlinearity in the pressure response is caused by three
     factors: second-order piezoresistive coefficients (Suzuki et al.,
     1985), membrane stress (large deflections) (Suzuki et al., 1985),
     and nonlinear boundary conditions reflecting the asymmetry
     of the micromachined structure. The device shown in Figure 5.21
     has a measured pressure nonlinearity of less than 0.1 percent

(gage and backward-gage configurations) for any diaphragm
thickness between 35 and 150 microns and a design rule of
20 mV/V full-scale output. At 35 microns, the deflection amounts
to 8.5 percent of the diaphragm thickness, and the maximum
strain is 40 ppm. At 150 microns and a reduced output of 10
mV/V, the deflection is 6.4 microns, the strain is 210 ppm,
and the measured pressure nonlinearity is 0.2 percent in the
gage configuration, but about 2 percent in the backward-gage
configuration. Quadratic piezoresistive coefficients have been
implemented in this simulation program. The above example
demonstrates the need to deal with the asymmetry of the micromachined structure. The series expansion fitted to the results
of a finite-element analysis has the flexibility to implement these
corrections (Lee and Wise, 1982).

To summarize, the method presented allows the transfer of
results from a mainframe to a simulation program implemented on
a personal computer. We have simulated a wide variety of different
chip designs and found excellent agreement with measured sensitivities.

# 6

# Sensor Specifications



## 6.1 INTRODUCTION

Piezoresistive transducers and sensors are specified in a data sheet.
One data sheet contains all the specifications of pressure ranges
and types of devices available in one series of sensors or transducers.
The data sheet is important, since the potential customer decides
which sensors to try in his or her system by comparing the specifications of different sensors. There is no standard for specifying
sensors, and each manufacturer writes specifications in its own
format. The potential customer has one basic question: Will this
sensor measure the pressure to 5 percent, 2 percent, or 0.5 percent
accuracy? Unfortunately, most data sheets do not specify the sensor
in terms of overall accuracy, but give individual parameters such
as linearity, repeatability, and hysteresis for each sensor. It is
left to the customer to look at each specification and arrive at an
overall accuracy. The basic specifications such as linearity, stability,
and temperature effects apply to sensors as well as transducers
and transmitters. We will now discuss the various specifications
and their impact on system accuracy.

## 6.2 SENSOR PERFORMANCE SPECIFICATIONS

Sensor specifications are given in a table format, with the following
headings:

*Minimum*:  A guaranteed limit; if the sensor measures less than
   this value, it can be returned during the warranty period.
*Maximum*:  A guaranteed limit; if the sensor measures more than
   this value, it can be returned during the warranty period.
*Typical*:  Not a guaranteed limit; may be an average value or just
   the value of the first devices.

If the data sheet has a heading called "Typical Specifications" and
the value is not followed by *minimum* or *maximum*, it is a typical
specification. Typical numbers should not be used as limits when
designing with a sensor, since there is no guarantee that the number
will not vary from purchase to purchase. Typical specifications give
general information about the sensor; for example, the bridge has
5-k$\Omega$ output impedance typical, as opposed to 2 k$\Omega$ or 500 $\Omega$.

Sensor specifications are given as absolute numbers—millivolts,
volts, psi, and ohms—or as a percent of full scale (FS). If the
parameter is expressed in the form % FSO, % FS, %/°C, or %, percent
of span or full scale is implied, not percent of reading.

The data sheet will list a set of reference or test conditions
under which the sensor is tested. The voltage or current supplied
to the bridge, a reference temperature, and pressure limits or
conversion factors may be included. The reference conditions may
be in a footnote or included as part of the table.

A list of materials used in the construction of the sensor or
a paragraph stating what medium is compatible with the sensor
should be included in the data sheet. Many sensors are compatible
only with clean, dry air, due to the construction of the chip or
package.

Maximum ratings are also given on the data sheet; these are
the maximum values that can be applied to the sensor without damage.
Included will be minimum and maximum operating temperature, minimum and maximum storage temperature, maximum supply voltage,
and maximum pressure that will not harm the device.


## 6.2.1  Accuracy

Accuracy is the combined effect of linearity, hysteresis, dead-band,
and repeatability errors. It is a figure of merit—how accurately
the sensor can measure an unknown pressure. When an accuracy
specification is given for a piezoresistive sensor, it is referring
to linearity and hysteresis and, maybe, repeatability. Notice that
these parameters are measured at the reference conditions and do
not include temperature effects. Temperature effects are added

to the basic accuracy numbers to give overall accuracy. The basic accuracy numbers are determined by the sensor, where the temperature effects can be compensated.

## 6.2.2  Pressure Range

The pressure range given on the data sheet is the pressure at which the device has been calibrated and tested. This pressure is a reference condition to which the customer and manufacturer can both test. Certain pressures have become standards with most manufacturers; 1 psi, 2 psi, 5 psi, 15 psi, 30 psi, 50 psi, 100 psi, 1000 psi, 3000 psi, and 5000 psi are common.

Most data sheets are not clear on two points. First, will gage sensors measure pressures both greater and less than ambient pressure? Gage-sensor pressure ranges are generally positive numbers (0 to 15 psi); this is the pressure at which the device is tested, but not the only pressure it can measure. A sensor will measure any positive or negative pressure up to the specified overpressure. This is not true of a signal-conditioned (amplified) transducer that has a high-level output. A transducer output will be close to ground at 0 psig and will saturate if a negative pressure is applied. The sensor inside the transducer is measuring the pressure, but the amplifier is limited by the power supply.

The second point is what compromises have to be made to use the sensor at pressures other than the specified pressure. For example, if the application is to measure 20 psi, should a 15-psi sensor be overpressured to 20 psi, or should a 30-psi sensor be used? Both will measure the 20 psi, but some compromise will have to be made:

*15-psi Sensor*:  The 15-psi sensor will have more sensitivity (mV/psi) with the same excitation, the offset TC will decrease as a percent of full scale, the linearity will degrade from the 15-psi value, and the device will be operated close to the maximum overpressure (probably 30 psi).

*30-psi Sensor*:  The 30-psi sensor will have less sensitivity with the same excitation, the offset TC will degrade, the linearity will improve when operating at a pressure less than the calibration pressure, and the maximum overpressure will be significantly higher (probably 60 psi).

The sensor selection will depend on the application; usually the higher pressure sensor is selected.

### 6.2.3  Calibration Specifications

*Offset*

Two types of sensors are available: the basic sensor, which is
the sensing die as processed, or calibrated and compensated sensors.
The wafers are sawed into dies, the dies are packaged and tested,
and the part is sold or rejected. The offset can be large, sometimes
almost as large as the span. Offset will be specified in millivolts
at the reference conditions. The offset will change when the sensor
is temperature-compensated, so the initial number is not a concern,
as long as it is not too large.

If the sensor is internally temperature-compensated, the offset
will be calibrated at the same time; 1 mV and 5 mV are common
offsets for calibrated sensors. The ratio of offset to span is more
important than the absolute number. If the span is 150 mV, 1-mV
maximum is pretty good, but if the span is 25 mV, 1-mV maximum
is large.

*Span*

Most manufacturers specify span, which is gage factor times applied
pressure at the reference conditions. Some manufacturers will specify
sensitivity or gage factor as well; gage factor is specified in mV/V/psi.
A basic packaged sensor will have a large tolerance on span or gage
factor; a range of 2/1 is common. The span will also be decreased
when the sensor is temperature-compensated, so the absolute numbers
are not too important. The minimum number may be of interest to
ensure a minimum sensitivity in the application.

The span is calibrated when the sensor is temperature-compensated.
A tolerance of 1 to 5 percent is common for most calibrated and
compensated sensors. Figure 6.1 shows the effects of offset and
span variations.

### 6.2.4  Stability and Repeatability

Stability and repeatability define the same parameter: repeatability
is a short-term specification, and stability is a long-term specification.
They are both defined as how accurately a sensor will repeat a
pressure measurement at any pressure within the specified pressure
range at a temperature within the calibrated temperature range.
Between measurements, the sensor is subjected to both pressure
and temperature cycles. Repeatability after full-scale pressure and
temperature cycles can be used as a figure of merit for the sensor
(the smaller the repeatability number, the better overall performance
of the sensor).

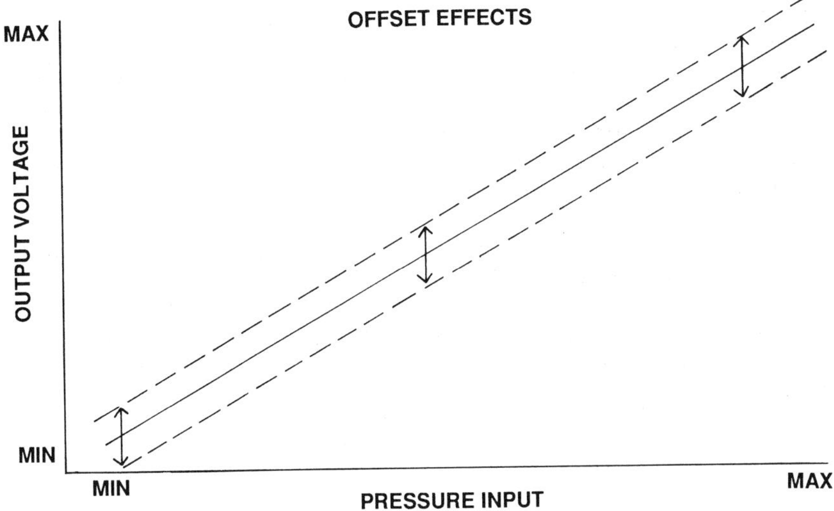


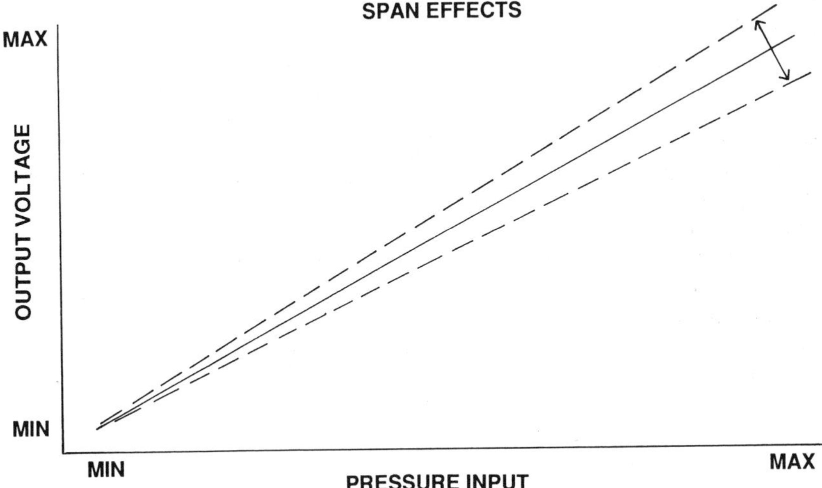


FIGURE 6.1  Offset and span effects.

*Repeatability*:   The maximum error of consecutive measurements
      at set reference conditions. Each measurement is followed by
      a number of full-scale pressure cycles and full-range temperature
      cycles. The next measurement is made at the conclusion of the
      temperature and pressure cycles.
*Stability*:   The maximum error of consecutive measurements at a
      set of reference conditions, 1 year between measurements.

These are very useful specifications when considering the overall
performance of the finished system. Here again, it must be a maximum
specification (a typical number should not be used). A sensor that
has a 0.5 percent maximum repeatability specification should not
be considered for a 1 percent overall accuracy system, since it
leaves 0.5 percent error for all remaining specifications.


## 6.2.5   Hysteresis

There are two forms of hysteresis, thermal and mechanical. Hysteresis
is defined as the maximum error after one specified cycle (pressure
or temperature); the cycle must end with a specified variable change
(ascending or descending variable).
      Due to the material used and the size of piezoresistive sensors,
mechanical hysteresis is extremely small, if not nonexistent. Hysteresis is determined by measuring the sensor at some reference condition,
then decreasing temperature or pressure to the lowest value, increasing to the highest value, and returning to the reference condition
(see Figure 6.2). The cycle must be the same for each measurement.
Care should be taken to ensure that the hysteresis being measured
is mechanical and not thermal. When the pressure is changed, the
density of the medium (generally air or nitrogen) changes, decreasing
the thermal resistance at the chip. This is especially true of high-
pressure sensors. The error will slowly decrease if it is thermally
induced.


## 6.2.6   Linearity

Linearity is defined as how closely the output of a sensor approximates a straight line when a linear pressure is applied. The nonlinearity is measured and expressed as linearity. The deviation
of the sensor calibration curve from a straight line is expressed
as a percent of full scale. Linearity is a function of the mechanical
chip design and the processing tolerances. Linearities of 0.1 to 1
percent are common in piezoresistive sensors.

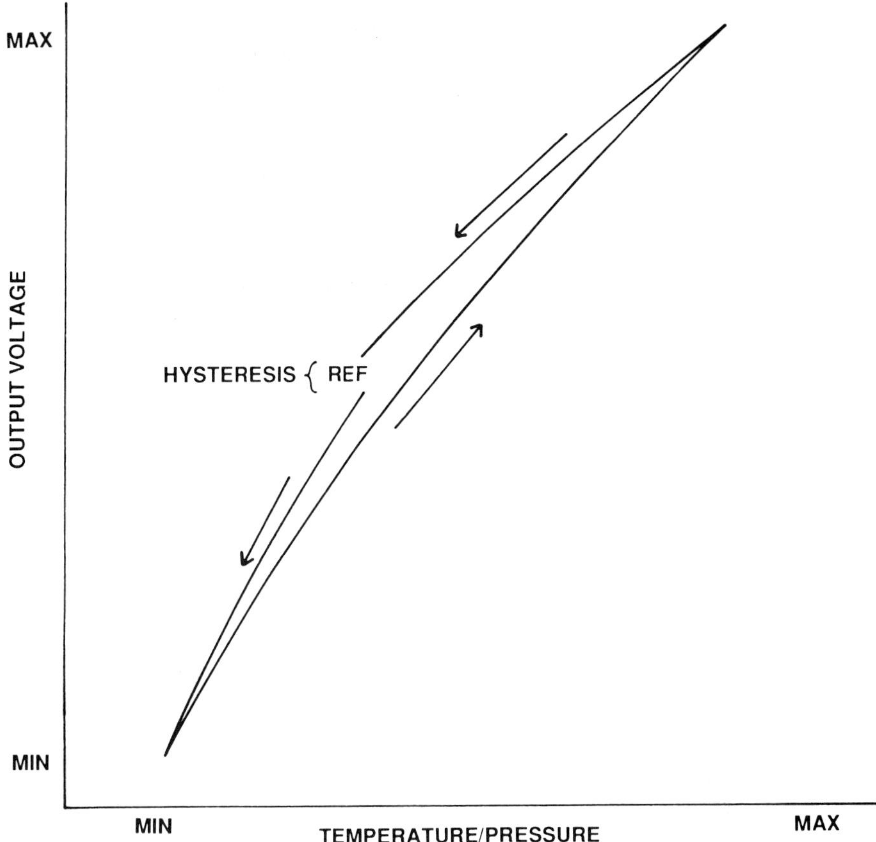



FIGURE 6.2  Hysteresis.

There are three accepted ways to express linearity: independent, terminal-based, and zero-based. If a data sheet does not specify a testing procedure for linearity, independent linearity is assumed.

*Independent Linearity*:  The maximum deviation of the calibration curve from a straight line so positioned as to minimize the maximum deviation (see Figure 6.3).

*Terminal-Based Linearity*:  The maximum deviation of the calibration curve from a straight line so positioned as to intersect the calibration curve at the minimum (zero) and maximum (full-scale) value (see Figure 6.4).

*Zero-Based Linearity*:   The maximum deviation of the calibration
   curve from a straight line so positioned as to intersect the
   calibration curve at the minimum value (zero) and to minimize
   the maximum deviation (see Figure 6.5).

Sensor data sheets specify independent linearity because it
results in the smallest number. If a system is calibrated at zero
and full scale, it is really using terminal-based linearity. If the
linearity error is a smooth curve, as shown in Figures 6.3 through
6.5, terminal-based linearity will be approximately twice the independent value. This should be taken into account when selecting
a sensor. The maximum independent linearity number should be
doubled if the system is going to be calibrated at zero and full
scale.

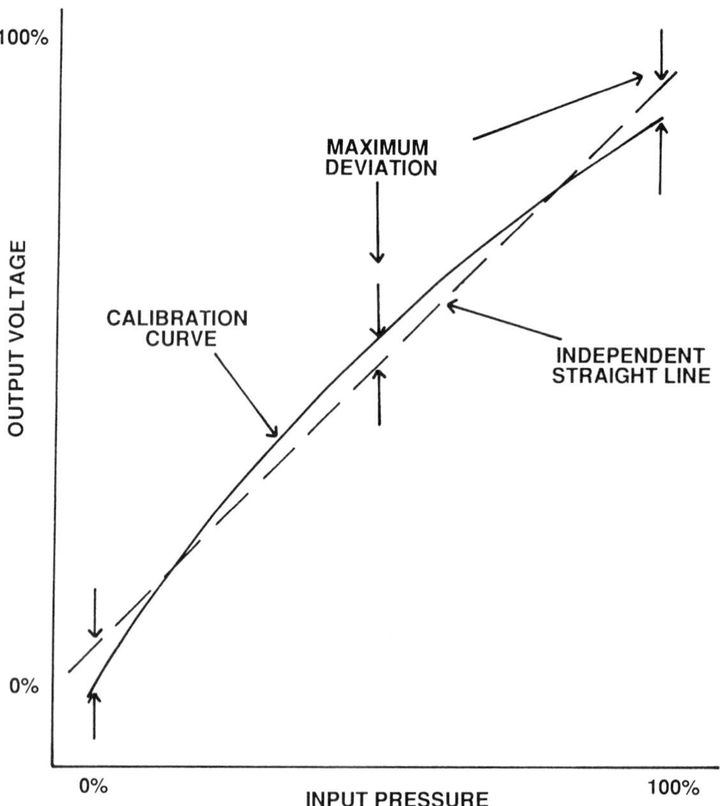


FIGURE 6.3   Independent linearity.

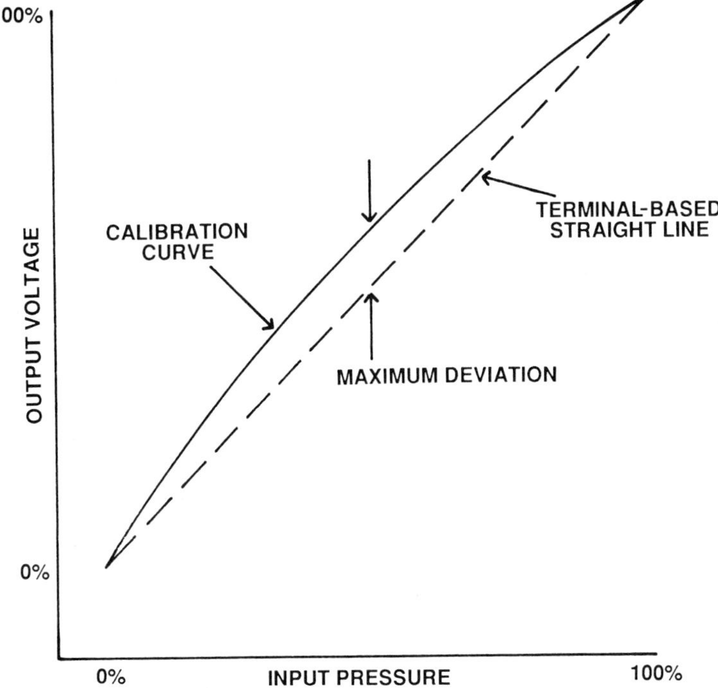


FIGURE 6.4   Terminal-based linearity.

Due to the media compatibility of the sensors, the pressure may be applied to the top side or the back side of the chip. With some low-pressure sensors, the difference in linearity error will be as much as 2/1, applying pressure to the front or back of the sensor. It should not be assumed that linearity will be the same under all conditions.

Linearity error is dependent on the applied pressure. The number specified on the data sheet is the maximum for the calibration pressure range. Figure 6.6 shows the linearity as a function of full-scale pressure. Full-scale pressure is defined as the maximum pressure that will be applied to the sensor—if 10 psi is applied to a 15-psi sensor, the full scale is 10 psi. If the linearity error is a smooth curve, the maximum nonlinearity is at 1/2 full scale. Terminal-based linearity is calculated:

$$V_L = V_1 + (V_3 - V_1/2) - V_2$$

where $V_L$ = nonlinearity in mV

$V_1$ = 0 pressure reading in mV (offset)

$V_2$ = half-scale reading in mV

$V_3$ = full-scale reading in mV

Nonlinearity as a percent of full scale is

Linearity % FS = $(V_L/V_3 - V_1) \times 100$

From Figure 6.6, the linearity error is not linear with respect
to applied pressure; that is, the 15-psi error is not twice the 7.5-psi
error. However, the linearity is not a step function: If the sensor
is overpressured to 150 percent of specified pressure, the nonlinearity
may increase 50 percent from the value at the specified pressure.

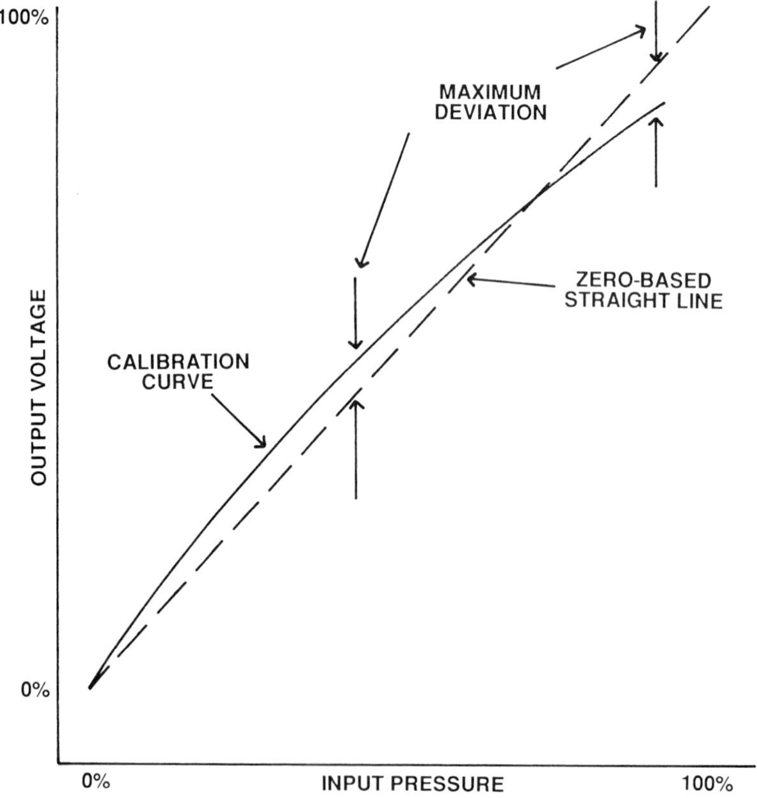


FIGURE 6.5   Zero-based linearity.

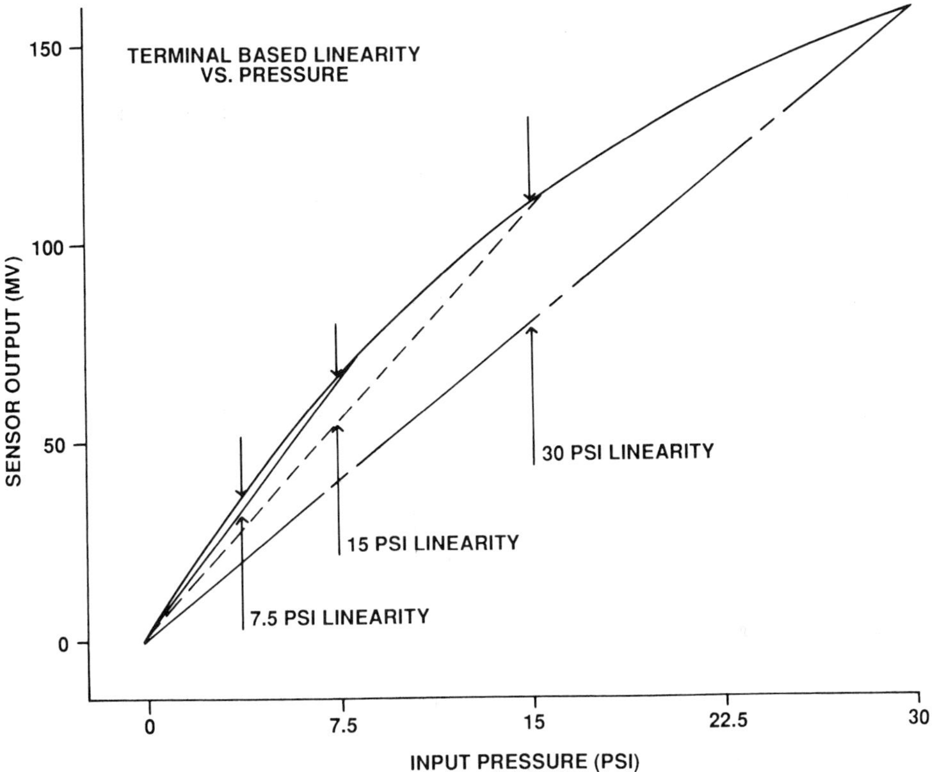



FIGURE 6.6   Linearity vs. full-scale pressure.

If the sensor has 0.6 percent nonlinearity at 15 psi, it will probably degrade to 0.7 or 0.8 percent at 20 psi (see Figure 6.6). The nonlinearity will increase more rapidly at pressures near maximum overpressure. The nonlinearity at pressures other than the calibrated pressure can be predicted by measuring a number of sensors to get the linearity/pressure curve, then using the maximum linearity at the specified pressure to calculate the linearity at the new working pressure. In a system where linearity is important, the designer can get a better idea of the system errors with this method than by using the specified error for all pressures.

Linearity error can also be decreased by calibrating at pressures other than zero and full scale. Calibrating at approximately 20 and 80 percent of full scale (independent linearity) will reduce the error to half. The disadvantage of this approach is that zero is not zero mV, but slightly negative voltage.

The linearity limits on the data sheet are independent linearity and apply to the rated pressure. The number on the data sheet is useful for comparing sensors and getting a general idea of the linearity range that can be expected (0.1 or 1 percent). If a more precise number is required, the independent linearity is converted to terminal-based linearity (independent times 2), and the pressure the sensor will be measuring has to be considered.

## 6.2.7  Temperature Coefficients

Piezoresistive sensors specify two temperature ranges: (1) the operating temperature range, which is the temperature range over which the sensor will operate (the sensor will function, but no error limits are given); and (2) the calibrated temperature range, which is the temeprature range over which the errors are calibrated and specified.

There is no standard as to how temperature errors will be specified. Some companies specify offset and span (or gage factor) separately, while others combine the errors and give a temperature-accuracy specification. Basic pressure sensors (no temperature compensation) will specify the range of parameters, and the customer will need to temperature-compensate the sensor. TC-GF and TCR (bridge-resistance TC) are the two parameters generally specified. Offset TC has to be measured on each sensor and would be very difficult to specify other than as a typical number. Sensors that include temperature compensation will specify temperature effects as percent full scale over a specified temperature range.

*Offset TC*  Because a number of factors will affect offset (for example, resistor tracking, stresses from the mount and package, and chip stresses that have temperature coefficients), each sensor must be measured over temperature to determine the sign and magnitude of the compensation. The pressure range of the sensor will determine which input (mechanical or electrical) will have the most effect. Low-pressure chips are affected more by the stress inputs, while high-pressure chips are affected by resistor tracking. The temperature-compensated sensors specify a maximum error over a temperature range ($\pm$1 percent full scale) or a maximum slope ($\pm$0.04 percent/$^\circ$C). Figure 6.7 shows the curves for the two specifications. The offset TC usually will not be a straight line; the curve will have a slight bow depending on how closely the offset TC matches the compensation circuit.

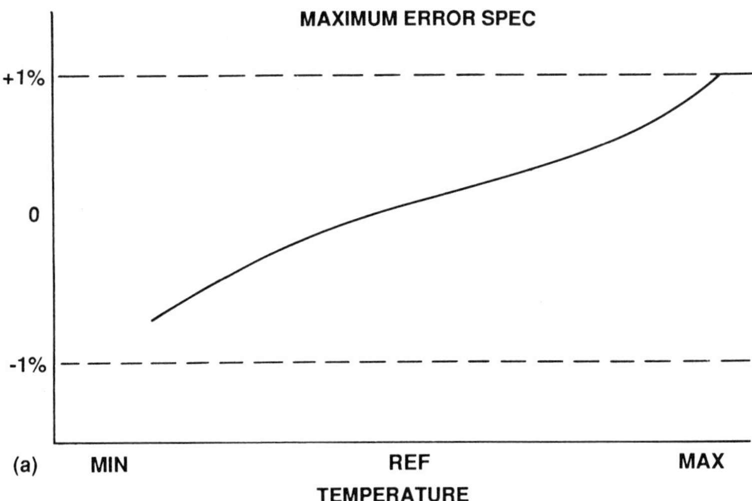



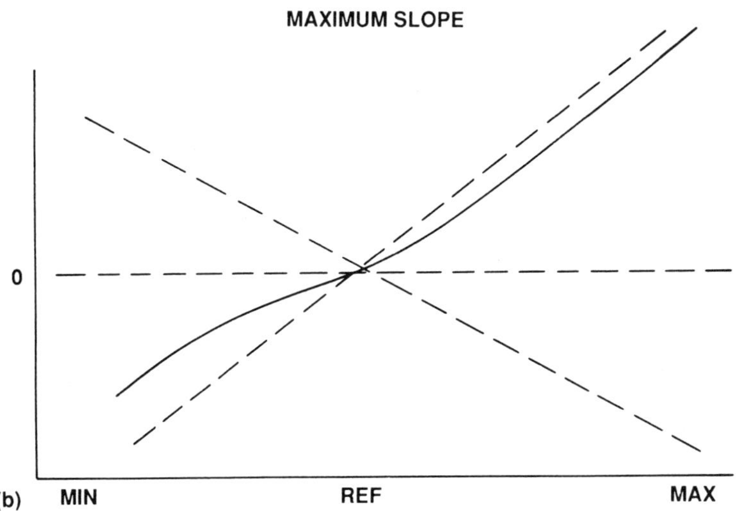



FIGURE 6.7   Offset TC.

*Span TC*:  Span TC is much more predictable than is offset TC.
The gage factor TC will always be negative and some nominal values
plus or minus a tolerance.  Most piezoresistive sensors have TC-GFs
of 1900 to 2400 ppm and are negative.  If a compensation circuit is
designed with a positive 2150 ppm, the span will be compensated to
within ±250 ppm, or ±0.025 percent/°C.

*Combined TC*  The combined TC specification is the most meaningful
to the user; it is generally in the form ±0.05 percent/°C.  This specification is an accuracy specification over temperature; for example, at
10°C above or below the reference temperature (25°C), the error will
be ±0.5 percent maximum.  To get an equivalent specification from the
individual specifications, the worst-case errors have to be added together, since the two errors could have opposite signs and cancel or
could have the same sign and add.

## 6.2.8  Bridge Impedance

Bridge impedance is a concern if a low-power system is being designed or if the sensor is driven by constant current. A sensor
that is powered from a voltage source will compensate for the input
impedance of the sensor. The excitation current multiplied by the
bridge impedance determines the voltage compliance that the designer
must allow for when designing the constant current source. The
output impedance may also be specified, but this is generally just
for information.

## 6.2.9  Package

The data sheet is the only source for package specifications. Package
dimensions, tolerances, and options are given in the data sheet.
If the sensor has a pipe fitting for a pressure port, the size and
type of thread are given, as well as any optional threads or package
material available. If the sensor is to be mounted on a printed circuit
board, the physical dimensions of the overall package, pressure
ports, electrical connections, and mounting holes are given. The
wire color code or pin function and electrical connection information
is also included in this section.

## 6.2.10  Optional Specifications

*Response Time*

Some manufacturers specify a response time or self-resonant frequency. Piezoresistive sensors will have a very high self-resonant

frequency because of their small size and single-piece silicon construction. Self-resonant frequencies of 30 kHz to 80 kHz are common. The plumbing to the sensor or the package design will generally determine the frequency response of the system; the sensor is much faster than most pressure changes. Response time is a difficult measurement; if gases are used, ruptured diaphragms are generally employed to measure response. The equivalent of a stiff balloon is inflated until it ruptures, and the sensor measures the shock wave. Another test is to hydraulically couple the shock wave to the sensor using a light oil or other inert liquid.

*Noise*

The one source of error found in all systems is noise. The higher the noise, the lower the resolution and accuracy of a system. Any system that requires high resolution will have significant noise problems. High-resolution designs require a familiarity with noise, its characteristics, and its sources. Low-frequency noise is the primary concern, since high-frequency noise can be filtered out in the electronics.

*Noise Sources*   Noise originates in the circuit components or is coupled into the circuit from external sources. Anything that generates an arc or spark generates wide-band noise; brushes in motors, spark plugs in cars, and fluorescent lamps all generate noise. Commercial radio stations and long-range radar around airports generate high-power interference signals that will be coupled into the high-gain circuitry.

The other noise sources are the components used in the circuit. Every electrical component whose temperature is above absolute zero (-273°C) generates small random-noise voltages and currents. This noise is referred to as *thermal noise*, because the motion of electrons in a conductor increases with temperature. Thermal noise is also called *resistance noise* or *Johnson noise*. The formula for the thermal noise developed across a resistor is

$$E_N^2 = 4kTBR$$

where $E_N$ = thermal noise voltage

$\quad\quad$ T = absolute temperature (degrees Kelvin)

$\quad\quad$ B = frequency bandwidth (Hz)

$\quad\quad$ R = resistance (ohms)

$\quad\quad$ k = Boltzmann's constant ($1.38 \times 10^{-23}$ joules/K)

From the equation, the noise is a function of temperature, frequency bandwidth, and bridge impedance. At 25°C, the noise equation reduces to

$$E_N = 1.28 \times 10^{-10} \, (BR)^{1/2}$$

If we assume a bridge impedance of 5 k$\Omega$ and a bandwidth of 100 Hz, a thermal noise voltage of $9.07 \times 10^{-8}$ V (0.09 $\mu$V) can be expected. This would represent the background noise for the ideal sensor measured under ideal conditions, not the total noise generated by the sensor.

Another natural noise source in a circuit component is *shot noise*. Shot noise is the small fluctuations in the DC current and is defined as

$$I_{SN} = (2eIB)^{1/2}$$

where $I_{SN}$ = shot-noise current (amps rms)

$e$ = charge on an electron ($1.6 \times 10^{-19}$ coulomb)

$I$ = direct current flowing (amps)

$B$ = bandwidth of the system (Hz)

Substituting the value for e, the equation becomes

$$I_{SN} = 5.65 \times 10^{-10} \, (IB)^{1/2}$$

Both thermal noise and shot noise are dependent on bandwidth and on impedance, in that if the impedance is low, the current is generally high. However, shot noise is not affected by temperature.

*Signal-to-Noise Ratio*   The signal-to-noise ratio of a system determines the maximum sensitivity possible. The ratio of signal power to noise power is generally used:

$$SNR = P_S/P_N = E_S^2/E_N^2$$

This ratio can be expressed as a ratio or in decibels:

$$SNR = 10 \log_{10} \, (P_S/P_N)$$

A ratio of 10 is 10 dB, 100 is 20 dB, and 1000 is 30 dB. The signal-to-noise ratio is a figure of merit—it is measured under ideal conditions and may or may not be close to the noise in the application. It is useful when selecting components for the system.

*Noise Measurements*   Noise cannot be eliminated, but it can be reduced with component selection and circuit layout. There are a number of ways to measure noise. A spectrum analyzer can be used, which will measure the noise voltage at each frequency.

Spectrum analyzers are expensive and are used in engineering to do some evaluations, but are not used in production. The most effective way to measure noise is to use the finished circuit and do a production noise test. Another way to measure noise is to use a strip-chart recorder or x/y recorder; these recorders are used to measure low-frequency noise (DC to 10 Hz). The recorder also gives a permanent record of the test. To measure noise in a sensor, a circuit has to be designed that has an order of magnitude less noise than the sensor to be measured. The following factors have a direct effect on the system noise.

Power supplies
Group loops
Lead lengths
Component values
Amplifiers
Filters

Power-supply regulators can generate significant amounts of noise; 60-Hz feedthrough and shot noise are generally present. Batteries can be used to test individual components. Some high-value tantalum capacitors will generate large amounts of noise. A large tantalum capacitor should not be placed at the input of a high-gain stage. Ceramic, mylar, or paper should be used instead.

Ground loops are generated when a component has more than one path to the power-supply ground. When the circuit or printed circuit board is laid out, the ground should resemble the spokes of a wheel—the center or hub is the power supply, with individual spokes or traces radiating out to the circuit components or groups of components.

Long leads and component values can pick up external noise signals. A long lead or trace at a high-impedance point in the circuit can act as an antenna. Summing junctions on operational amplifiers and other high-impedance points should have minimum lead length.

The frequency bandwidth will have a direct effect on the noise in a circuit. A majority of pressure-sensor applications are very slow and do not require a fast measurement. A bandwidth in the tens of hertz will do in most applications. The bandwidth can be reduced with simple resistance-capacitance filters or more sophisticated active filters.

The operational amplifier used in the circuit is just as important as the sensor, since the noise generated at the amplifier input gets amplified with the sensor output. Looking at the amplifier output, it would be impossible to distinguish the sensor noise from the amplifier noise. Some general-purpose amplifiers can generate

many times the noise generated at the sensor. In low-resolution
applications, these amplifiers are adequate, but high-resolution
applications require better amplifiers. If an amplifier does not specify
noise, it is most likely pretty high. Some general-purpose amplifiers
will specify noise and, if noise is a major concern, special low-noise
amplifiers are available. If a special low-noise amplifier is not used,
the amplifier will generate more noise than does the sensor. To
determine the noise source, the noise is measured—the sensor output
is shorted and the noise is measured again. If the noise is not
considerably reduced, the amplifier is the main noise source.

*Transducers*   Signal-conditioned and amplified transducers will have
significantly more noise than will sensors. The equivalent noise
in a sensor is about $2 \times 10^{-4}$ psi (5-psi sensor), while an amplified
transducer has as much as $1 \times 10^{-3}$ psi (5 psi, 1-V/psi transducer).
Both devices use the same sensor die; the signal-to-noise ratio
of the transducer is reduced by a factor of five by the amplifier.

   Since the noise generated in the sensor is electrical—and is
affected by impedance, current, and temperature—the sensor noise
will be independent of pressure range. If the perfect amplifier
is used, the sensor noise will determine the maximum sensitivity
possible. Amplified transducer noise voltage will vary with pressure
range—the lower the pressure range, the more gain in the amplifier
stage (the amplifier being the main noise source). If noise is a
potential problem, a sensor and well-designed low-noise amplifier
stage should be used.


## 6.3  SPECIALS

Most manufacturers will make sensors to special specifications if
it can be done economically. A minimum order and minimum lot size
will have to be agreed upon, and the price will increase depending
on the modified specification. Some manufacturers will sell engineering—
that is, they will design a sensor to your specifications. They will
develop all the necessary processes to manufacture the sensor.
The customer generally pays a fixed amount for the development
and may or may not have the device manufactured at the same
company. Not all sensor companies are set up to sell engineering;
if the company is a high-volume producer of just a few devices,
it probably would not be interested.


### 6.3.1  Special Testing

The most common special is testing to limits other than standard
or testing a parameter that is not normally tested. For example, if

the specified span of an uncompensated sensor has a range of 50 to 200 mV but the customer needs sensors with a range of 75 to 125 mV, a special specification can be written and tested in production. These devices have to be marked and handled differently from the standard product. The price increase covers the special handling and yield of the tighter specifications (yield is the percent of units tested that meet the specifications).

Yield is very important to both the customer and the manufacturer. If the yield is too low, the price increases, and the manufacturer could end up overstocked with slow-moving parts. Most parameters have a normal distribution curve (Figure 6.8); the majority of the units will be grouped around a mean value, with the number of units dropping off to zero at some minimum and maximum value. The specified limits will include this distribution plus any shifting of the mean value due to process variables. The best yield will be at the mean value. If only 5 units out of 100 meet the specification, 95 units go into stock; at this yield, 2000

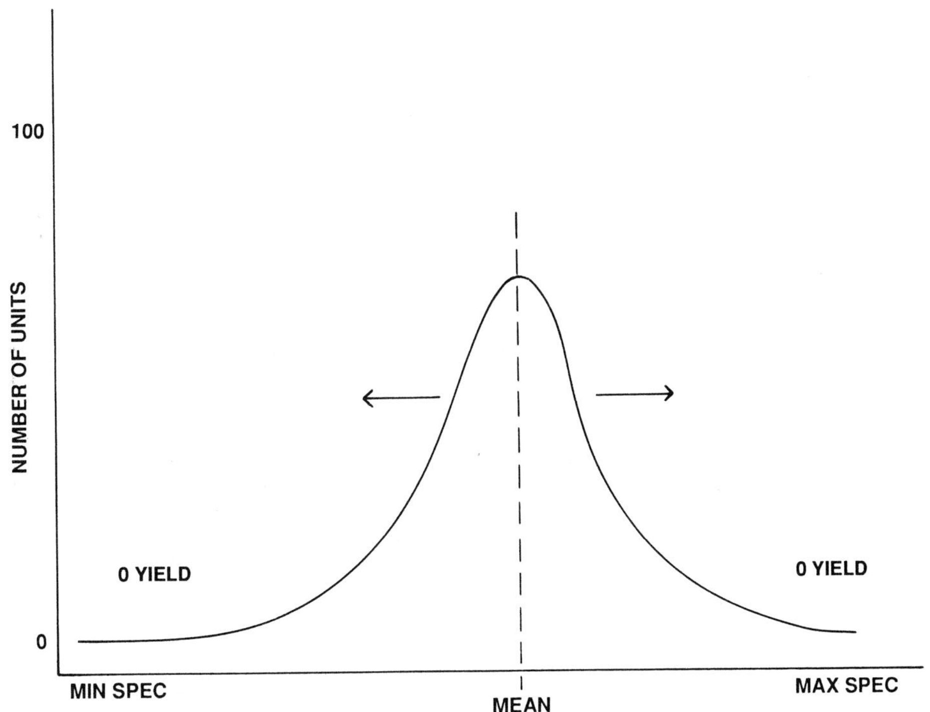


FIGURE 6.8   Distribution curve.

devices would have to be tested to get 100 parts. If the manufacturer
and customer get together before the specification is written, look
at the distribution of the sensor, and adjust the limits for a more
favorable yield, both benefit.

Another special test is to increase the limits on some tests that
the customer does not need. If the manufacturer is getting a low
yield from one specification, he may agree to sell the fallout at
a reduced price. This works only if the basic sensor is a high-
volume part and there is plenty of fallout to meet the other customers'
needs. These parts require special handling and marking to ensure
that they are not sold as standard product.

Noise is another specification that is tested as a special. Noise
is very difficult to test, and the customer and manufacturer have
to set up a test that is reproducible and gives the customer units
that work in his or her system. Some customers will give the sensor
manufacturer a breadboard of their system, and the manufacturer
uses this circuit as a fixture to do the noise test. Or, each sensor
could be run on a strip-chart recorder and the noise plot given
to the customer with the sensor.

A customer may request that read and recorded data be shipped
with each sensor. Most sensor manufacturers are not set up to
read and record, due to batch processing and testing. Each sensor
would have to be serialized and tracked through the final testing.
This could be very expensive, if it were agreed to at all.


## 6.3.2 Special Trimming

Low-cost transducers are generally hybrids. A thick-film circuit
that has all the resistors and thermistors is used to signal-condition
the sensor. The sensor may or may not be mounted on the ceramic.
The trend is to have the sensor in a separate package to isolate
the medium from the signal-conditioning circuits. Temperature-
compensated and calibrated sensors are also available using the
same technique. After the device is assembled, it is tested and
the data are serialized and stored. A computer-controlled laser
reads the data and trims the thick-film resistors to temperature-
compensate and calibrate the sensor. Each family of sensors will
have a different trim program and test procedure. If a customer
requires a special trim, a modified trim program has to be written
and the devices need special marking and handling.

A customer may request an offset voltage and sensitivity that
will be compatible with the analog-to-digital (A/D) converter in
his or her circuit or will drive a special recorder. Special temperature testing is common for compensated sensors and transducers.
They may require special trimming or testing to different limits.

### 6.3.3  Special Package

Some customers cannot use the standard sensor package for one reason or another. In these cases, customers generally have the package designed and submit the outline of the package they would like to purchase. Custom packaging can be handled in one of two ways: The customer provides the piece parts and the manufacturer assembles and tests the sensor, or the manufacturer designs the piece parts and supplies the finished product to the customer. Custom packages are costly and time-consuming; not only does the package have to be designed, but test fixtures and laser-trim fixtures have to be machined to test and calibrate the devices. Most manufacturers try to leave any special packaging to the customer.

### 6.3.4  Military Specifications

To sell sensors to the military or government agencies requries special testing and documentation of each sensor. Most sensor manufacturers are not set up to sell to the military, and those that are have set up part of the company to deal exclusively with these accounts. Some of the tests involved are:

*Source Inspection*:  The government agency will send inspectors to the manufacturer's plant to ensure that all documentation is in order. Each step in the assembly procedure and test must be documented and on file. All test equipment must have up-to-date calibration stickers, and all quality-control documentation is checked.

*Burn-in*:  After the devices have been assembled and calibrated, they are placed in an oven and power is applied for a period of time (typically 168 hours).

*Vibration*:  Each sensor is placed on a shake table and vibrated in each of the three axes.

*Shock*:  Each sensor is dropped on a special surface from a specified height a number of times.

*Temperature/Humidity*:  The sensors are placed in a special oven capable of controlling both temperature and humidity. The oven is programmed for a temperature/humidity profile.

*Traceability*:  Each component used in the sensor has to be traceable back to its origin

Military testing requires a lot of special equipment and personnel to do the extra paperwork. For some applications, military agencies will purchase parts with minimum testing or no special testing and do the tests on the finished product instead of on each individual component. This reduces the price of each component significantly.

# 7
# Signal Conditioning

## 7.1 INTRODUCTION

To better understand sensor signal conditioning, we will discuss
the components used to condition sensors. With the advancements
of the past decade in the electronics industry, precision components
are available at very reasonable prices. Operational amplifiers and
microprocessors have made the largest advances in performance
specifications. The temperature-compensated sensors have simplified
the signal conditioning necessary to use the sensors. Using some
basic guidelines, anyone can design and build sensor signal-conditioning
circuits. Amplification and calibration are the only things left to
the user. The basic question of whether customers should buy
a signal-conditioned transducer or purchase a temperature-compensated
sensor and do their own signal conditioning depends on the number
of sensors that will be used and the type of application. If the
customer is buying a finished transducer and adding amplifiers
to level-shift or change the gain, a temperature-compensated sensor
should be investigated. The finished transducer makes sense if
there are no printed circuit boards in the system or if the transducer is mounted some distance from the electronics. In this chapter,
we will discuss some of the basic components used to condition
sensors, operational amplifiers, instrumentation amplifiers, references,
and passive components.

## 7.2  OPERATIONAL AMPLIFIERS

The operational amplifier is an extremely efficient and versatile device. Many analog transfer functions can be designed with a basic understanding of the negative-feedback technique used with operational amplifiers. What the operational amplifier can do is limited only by the imagination and ingenuity of the user.

The operational amplifier is a DC-voltage amplifier that has differential inputs and single-ended output. Figure 7.1 shows the basic block diagram of an operational amplifier. The differential input stage may be designed for a special function, such as high-input impedance (FET input), speed, large common-mode range, or low noise. The second stage is the gain stage; operational amplifiers have very high open-loop gain (80-100 dB). The higher the open-loop gain, the more accurate the closed-loop operation. The third stage is the output stage, which buffers the gain stage from the load and converts the differential gain-stage output to a single-ended, high-power output.

In the beginning, operational amplifiers used vacuum tubes and were quite large. The next generation used discrete transistors, resistors, and capacitors. The first monolithic operational amplifier (an amplifier on one chip) was the μa709, which cost about $100. Over the years, the technology has made breakthrough after break-through; now hundreds of different amplifiers are available, ranging from low-cost, general-purpose amplifiers to special-function amplifiers. Most special-function amplifiers have one characteristic such as high input impedance (low bias current), low noise, high slew rate, or low power. These amplifiers cost a little more, but are well worth the price. Amplifier packaging has also made breakthroughs: Up to four amplifiers are available in one 14-pin DIP (dual in-line package). Today, operational amplifiers are used as transistors

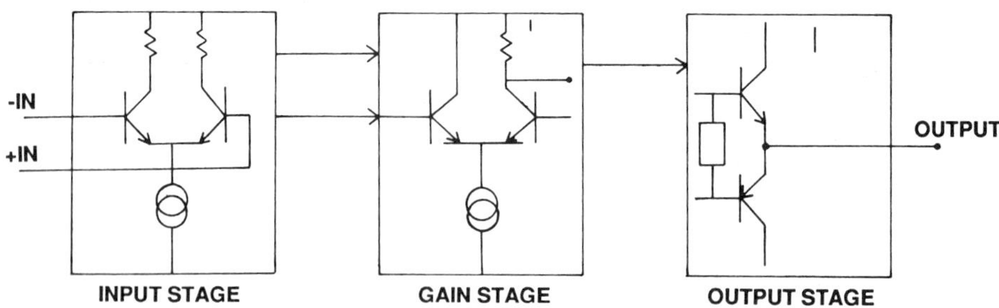


FIGURE 7.1  Operational amplifier.

were used 20 years ago. In quantity, general-purpose amplifiers can be purchased for 20¢ to 40¢ apiece.


### 7.2.1  Feedback

The operational amplifier is a basic building block that can be used to design many analog functions by designing the feedback circuit around the operational amplifier. Positive feedback is used to create hysteresis in comparator circuits and to make oscillators. An oscillator has positive feedback and a gain of at least 1. Negative feedback is used to set the gain of the stage precisely.

The operational amplifier has differential inputs: (1) the negative input—if a positive voltage is applied to this input, the output will swing negative; and (2) the positive input—if a positive voltage is applied to this input, the output will swing positive.

Two things need to be remembered about amplifier inputs: First, due to the high input impedance of the amplifier, a very small current flows in the inputs (nanoamps—$1 \times 10^{-9}$ amp); and, second, the inputs will be forced to the same bias voltage by negative feedback. The negative input is sometimes referred to as the *summing point* or *summing junction*. Electrically, this input will look like a pseudo-ground. The voltage at the negative input is forced to the same potential as the positive input by the negative feedback.

The two basic feedback configurations are shown in Figure 7.2. Figure 7.2a shows the noninverting configuration, where the input is connected to the positive input. The voltage gain is determined by the ratio of the two feedback resistors:

$$E_0 = 1 + (R_0 / R_1) \tag{7.1}$$

and the output is in phase with the input. This configuration has high input impedance (megohms).

Figure 7.2b shows the inverting configuration, where the positive input is grounded, the input voltage is applied to the input resistor $R_1$, and the gain is determined by the ratio of the feedback resistors:

$$E_0 = R_0 / R_1 \tag{7.2}$$

The output is 180 degrees out of phase with the input. The tolerance of the two feedback resistors determines the tolerance of the amplifier gain. The input impedance of the inverting amplifier is equal to the value of $R_1$. In both configurations, the voltage at the inputs ($E_2$) is equal to the input offset voltage of the amplifier. These are the two classic configurations; many different circuits can be designed by modifying these basic circuits.

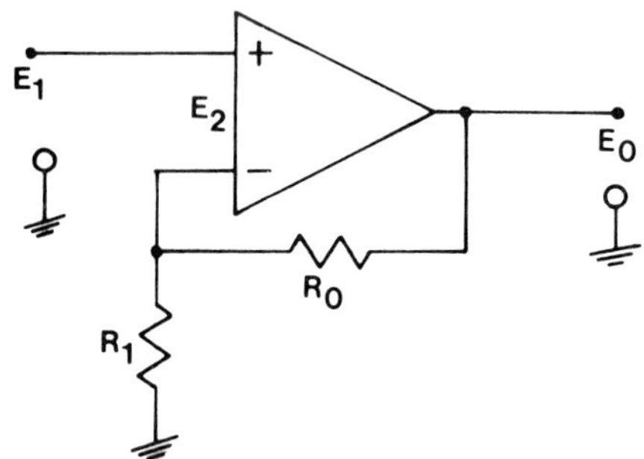

**NONINVERTING CONFIGURATION**

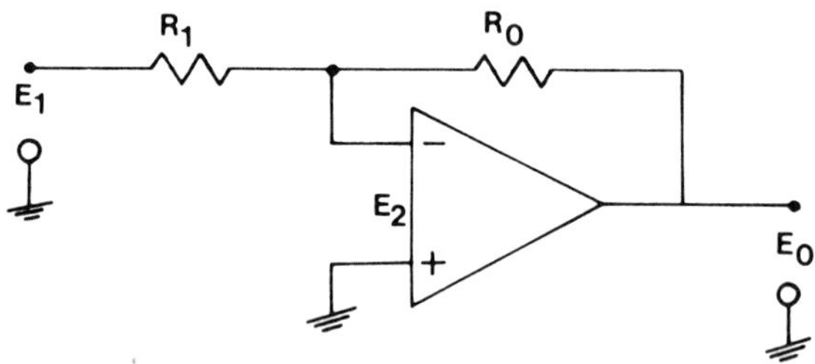

**INVERTING CONFIGURATION**

FIGURE 7.2   Basic feedback configurations.

Due to the high open-loop gain at low frequencies (assume 100,000 to 100 dB open-loop gain), a very small voltage change at the input will saturate the output if negative feedback is not used. If the output saturates at ±10 V, the maximum input swing is 20 V/100,000, or $2 \times 10^{-4}$ V. A voltage change of ±200 μV will saturate the amplifier. If the output voltage is fed back to the negative input through a voltage divider, the amplifier forces the two inputs to the same voltage, and the output changes as the ratio of the voltage divider. Positive feedback (hysteresis) is used with comparators to give clean transitions from one state to another. If no hysteresis is used around a comparator, the output will oscillate at the switch point, due to the noise voltage. If a small signal is coupled back to the positive input through a voltage divider, the output will not change state until the voltage that has been fed back is exceeded. The positive feedback is generally designed to be just larger than the maximum noise voltage.

Positive feedback is also used when designing oscillators. If a specific frequency is selected and fed back to the input, the circuit will oscillate at that frequency. Feedback-circuit designs may vary, but they all use positive feedback.

### 7.2.2 Summing Amplifier

The operational amplifier can also be used to add signals together (see Figure 7.3).

$$V_1 = V_2 = \text{ground potential}$$

$$-I_0 = I_1 + I_2 + I_3 \tag{7.3}$$

The current $I_0$ is

$$I_0 = -E_0/R_0 = E_1/R_1 + E_2/R_2 + E_3/R_3 \tag{7.4}$$

Each input, $E_n$, is multiplied by a factor, $-R_0/R_n$, before summing. Each input can be amplified or attenuated independently. This is very convenient if the signals need to be adjusted before summing.

The output of a sensor can be amplified using an instrumentation amplifier stage followed by another amplifier. The second amplifier can be used to sum a temperature-correction signal to the sensor

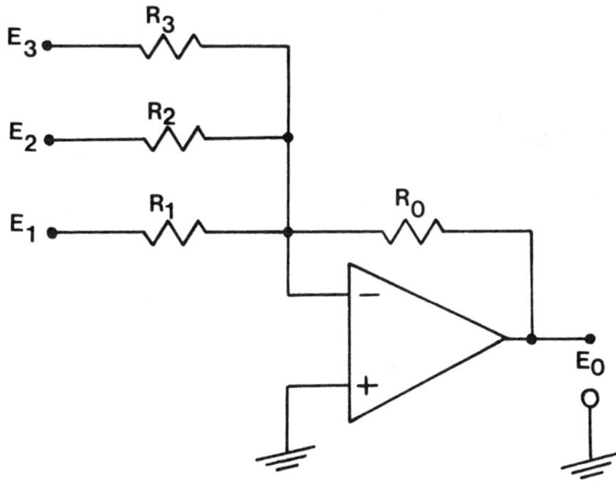


FIGURE 7.3  Summing amplifier.

output. Each input can be amplified or attenuated by the ratio $R_0/R_{in}$. This can be very useful when designing compensation circuits; for example, an offset-TC signal can be summed with the sensor output. The sensor signal may have a gain of 1, while the temperature-correction signal is amplified to compensate the TC error at the output.

### 7.2.3  Frequency Response and Stability

The frequency of an operational amplifier is a function of the amplifier frequency compensation. When the amplifier is designed, a frequency-compensation network is built into the circuit. With no compensation, all operational amplifiers would oscillate when feedback is applied. The compensation decreases the open-loop gain to ensure that the amplifier has a gain of less than 1 before the phase of the output shifts and becomes positive feedback.

Most general-purpose operational amplifiers available today are internally compensated; that is, no capacitors have to be added to the amplifier to *roll it off*. The compensation capacitors are processed on the silicon chip. Some special-purpose amplifiers still have external frequency compensation. If the speed or bandwidth is adjustable, the external compensation is modified by changing the compensation capacitor and resistor. Figure 7.4 shows the Bode plot of a typical amplifier. An RC network is added to the amplifier to ensure that the gain is less than 1 when the phase of the output becomes in phase with the input signal. By definition, an oscillator must have a gain of at least 1 and the output in phase with the input. If the phase of the amplifier output is allowed to become positive feedback before the gain is less than 1, the circuit will oscillate. To ensure a stable circuit, and amplifier with a guaranteed phase margin should be used. The phase margin is an indication of how stable the amplifier will be at low gains. External capacitance and inductance will affect the phase margin at low gains. Capacitive loads at the amplifier output and long power-supply leads can make some amplifiers oscillate. In dealing with sensors, all the amplifiers are fast enough to reproduce the pressure signal. The only concern is stability; if the circuit has long power-supply leads, the inductance on the wire can cause the amplifier to oscillate. It is a good practice to bypass the power-supply leads as close to the amplifier as possible, especially if no phase-margin specification is given for the amplifier. Phase margin is critical in voltage follower circuits or circuits with low gains (1 to $\pm0$). Phase margin in circuits with high gains ($>10$) is not too critical.

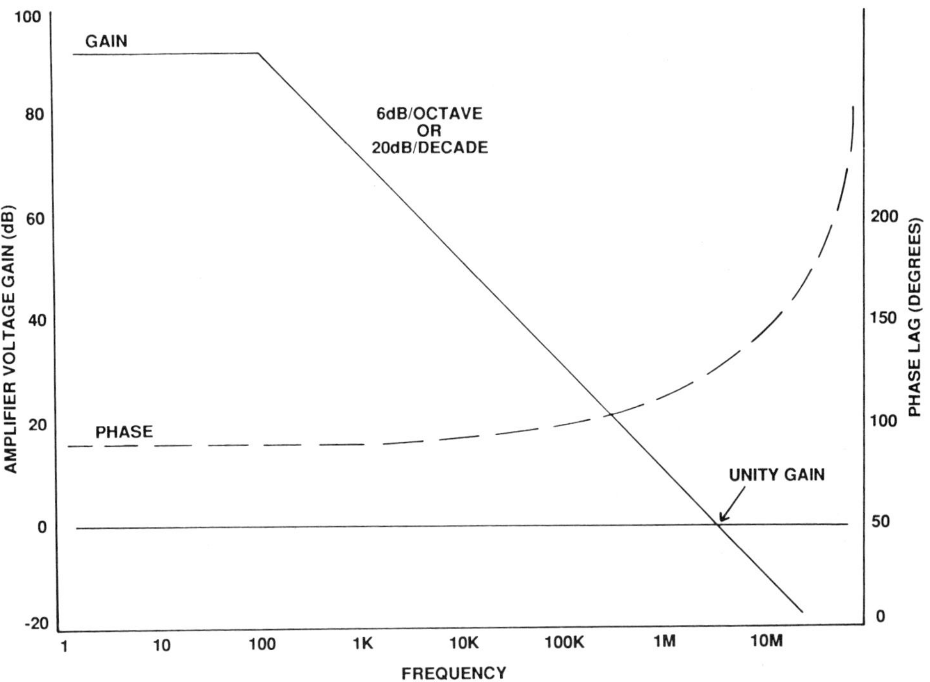



FIGURE 7.4   Amplifier Bode plot.

### 7.2.4  Operational Amplifier Specifications

Operational amplifiers are specified by placing limits on each amplifier
parameter. There are so many amplifiers available that some discussion
of the amplifier specifications that are of concern to sensor users
is appropriate.

*Maximum Ratings*

Each amplifier lists the maximum power-supply voltage, differential-
input voltage, storage-temperature range, and output short-circuit
duration. If these ratings are exceeded, the amplifier could fail
or degrade. These are maximum limits, not the values that the
amplifier is intended to operate with.

*Power Supply*

The range of power-supply voltages over which the amplifier will
work is important to the sensor user. The standard supply voltage

is ±15 V DC; some amplifiers need the full 30 V to operate, while others operate from a single supply. If a single supply is being used, the operating voltage range of the amplifier should be checked. Some of the new designs will operate from a single supply of 3.0 to 30.0 V.

### Power Dissipation

In some designs, power dissipation can be very critical. Battery-operated systems that have to run for long periods of time need minimum power consumption. Some amplifiers are designed to operate from supply currents of 200 to 500 $\mu$A.

### Input Offset Voltage

Input offset voltage is the offset or mismatch of the input transistors or field-effect transistors. As a general rule, the more close the input transistors are matched, the better the overall performance over temperature and time. Offset voltage is important when dealing with sensor signal-conditioning circuits. Spending extra money for a sensor that has ±1-mV offset calibration and amplifying the output with an amplifier that has ±8-mV maximum offset voltage would not be practical. A general-purpose amplifier could have an offset voltage as high as ±10 mV, but ±0.5-mV to ±2-mV input offset voltage is available at a reasonable cost.

### Input Offset Current

The input offset current is the difference in the two input currents. Standard amplifiers have offset currents in the range of 0.5 to 50 nA (nA = 1 $\times$ 10$^{-9}$ amp). Amplifiers with FET input stages have 1 to 100 picoamps (pA) of offset current. Input offset current can be important if the amplifier is used with high-impedance gain-setting networks.

### Input Bias Current

Input bias current is the average current in both inputs. The bias current is an order of magnitude larger than the offset current and, like the offset current, can generate error voltages in high-impedance circuits.

### Open-Loop Voltage Gain

The open-loop voltage gain is the differential voltage gain of the amplifier. It can be expressed in dB or as a ratio (V/mV). Gains of 10,000 to 100,000 are common for general-purpose amplifiers. Most sensor designs have fairly low amplifier gains, so the open-loop gain should not be a major problem.

## Common-Mode Voltage Range

Common-mode specifications apply to both inputs as opposed to the differential (difference in the two inputs). The common-mode voltage range is the range of voltage over which the amplifier inputs can be biased. If the amplifier is biased beyond these limits, the input stage becomes saturated and the amplifier cannot function. Most amplifiers operate from 1.5 volts above the negative supply and 2.0 volts below the positive supply. Some amplifiers operate with the inputs at the negative supply or ground potential, but not all the way to the positive supply. Common-mode voltage range is important to the sensor user. Current sources and offset-adjust circuits may need to operate close to the power-supply potential.

## Output Voltage Swing

The majority of the general-purpose amplifiers swing to within 2 V of the supply voltage. The NPN and PNP transistors in the output require approximately 1.5 V to operate. A few of the new designs saturate the PNP transistor, allowing the output to get within 150 mV of the negative supply. Some of the new CMOS (complementary metal oxide silicon) designs swing from supply to supply. This can be important when interfacing to some of the new analog-to-digital converters that have an input voltage range of 0 to 5 V and operate from a single 5-V supply.

## Temperature Effects

Input offset voltage, input bias current, and input offset current are specified over a particular temperature range. Amplifiers are available in three grades or classes: Commercial grade is specified over a temperature range of 0°C to 70°C, standard grade is specified from -25°C to 85°C, and military grade is specified from -55°C to 125°C.

## Noise

Most amplifiers do not specify noise. A graph may be included in the data sheet showing typical noise characteristics, but no limit is given. If a high-sensitivity circuit is being designed, an amplifier that specifies noise should be selected. Special low-noise amplifiers are also available for critical applications. Noise is a major problem when designing high-resolution sensor signal-conditioning circuits.

## Slew Rate

Slew rate defines the speed of the amplifier. It is generally given in V/μs. A general-purpose amplifier will have a slew rate of approximately 0.5 V/μs. High-speed amplifiers are available with slew rates

of 10 V/µs or more. General-purpose amplifiers are fast enough
for most sensor applications.

*Package*

Amplifiers are available in a variety of packages, depending on
the temperature range of the device. Commercial packages are plastic;
standard and military packages may be ceramic or metal. Single
and dual amplifiers are available in 8-pin dual in-line packages
or 8-pin TO-5 packages. Amplifiers are available with one, two,
or four amplifiers per package. The four-amplifier package can
decrease the package count if a large number of amplifiers are
used.


## 7.3  INSTRUMENTATION AMPLIFIERS

Instrumentation amplifiers are special-purpose differential amplifiers
that are characterized by

1.  Differential inputs
2.  High input impedance (many megohms)
3.  Reference input (used to level-shift the output)
4.  Single gain-adjust resistor

To get the high input impedance, the differential input signal is
applied directly to the input transistors, not through a resistive
input network. A reference input is provided on most instrumentation
amplifiers to level-shift or adjust the output offset voltage. (With
0.V differential applied to the amplifier inputs, the output voltage
will be equal to the reference voltage.) The differential inputs
and high input impedance make the instrumentation amplifier ideal
for amplifying temperature-compensated sensors, since most sensors
are passively compensated (resistors are used to parallel certain
bridge resistors). If an amplifier with low input impedance is used,
it will shunt the bridge output, changing the factory compensation.
Instrumentation amplifiers also have very good common-mode rejection.
Common-mode rejection is important, since all piezoresistive sensors
compensate the span temperature coefficient by changing the voltage
developed across the bridge.
   Instrumentation amplifiers can have active or passive feedback
networks, and the gain is adjusted by one resistor. Instrumentation
amplifiers can be purchased or designed into the signal-conditioning
electronics using operational amplifiers. Most instrumentation amplifiers available on the market have certain restrictions: They must
operate from ±15-V power supplies; the output swings ±10 V; and,
depending on the specifications, they can be expensive.

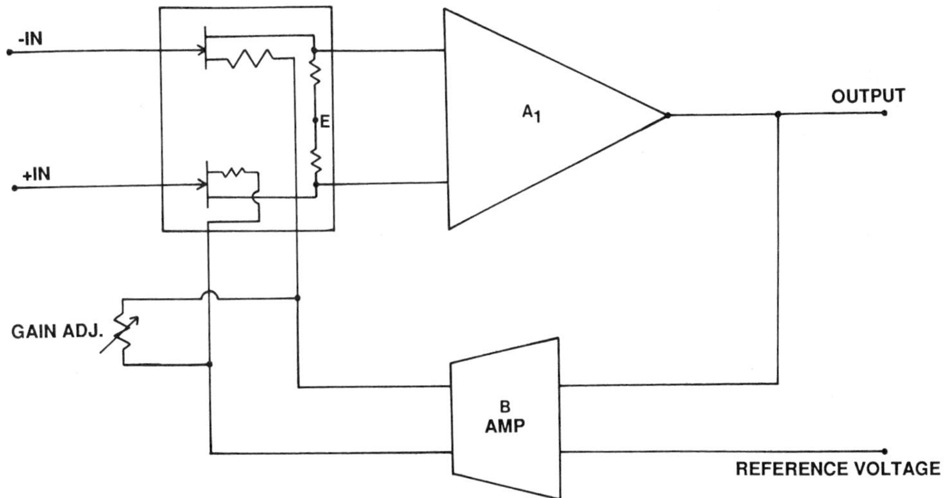

FIGURE 7.5   Active feedback instrumentation amplifier.

## 7.3.1   IC Instrumentation Amplifiers

The two most common packaged instrumentation amplifiers are integrated circuits and hybrids. The integrated circuit is on one chip, while the hybrid is constructed of individual amplifier chips and thick-film resistors that are adjusted by laser trimming.

There are many approaches to designing instrumentation amplifiers. Figure 7.5 shows a design using an active feedback amplifier. This design has true differential inputs and high input impedance due to the FET input stage. The amplifier has three stages: the input stage, the amplifier stage, and the $\beta$-amplifier stage, which senses the output voltage and the reference voltage and provides negative feedback to the input stage.

The input stage can be designed using transistors or FETs. The FET has lower bias current and higher input impedance, but more noise, than does the transistor input. The $\beta$-amplifier has differential current outputs that are out of phase with the input signal and are used to bias the input stage. The input stage sets the gain of the amplifier.

The amplifier stage can be a conventional operational amplifier that amplifies the differential output of the input stage and provides a high-power output signal to the load.

The $\beta$-amp or feedback amplifier sets the output voltage and provides negative feedback to the input stage. With 0 V applied

to the inputs, the β-amp forces the output to the same voltage
as the reference voltage. The reference input can be used to level-
shift the output or as a fine offset adjustment. The β-amp also
senses the difference in the output and reference input, amplifies
this signal, and converts it to a differential current. Shunting
the β-amp output sets the overall gain. If the gain resistor is opened,
the overall gain will be less than 1; if the gain resistor is shorted
(0 feedback), the amplifier will be open loop—maximum gain. The
gain is set by adjusting the value of the shunt resistor. The value
of the resistor is generally calculated.

The passive-feedback instrumentation designs use instrumentation
amplifiers and resistors arranged in such a manner as to have the
basic instrumentation-amplifier characteristics of high input impedance,
reference input, and one-resistor gain adjust.


### 7.3.2  Difference Amplifier

*Gain Calculations*

The simplest difference amplifier is shown in Figure 7.6. This circuit
is a combination of the inverting and noninverting operational amplifier configurations. The gain is given by

$$V_0 = V_2 (R_4/R_3 + R_4)(R_1 + R_2/R_1) - V_1 (R_2/R_1) + V_r \qquad (7.5)$$

If the resistor values are selected such that $R_4/R_3 = R_2/R_1$:

$$V_0 = R_2/R_1 (V_2 - V_1) + V_r \qquad (7.6)$$

If the differential input is zero ($V_1 = V_2$), the first term goes
to zero and the output is equal to $V_r$.

*Circuit Limitations*

This configuration has basic problems amplifying resistive-compensated
sensors. The input impedance is determined by the feedback resistors,
pairs of resistors have to be changed to change the amplifier gain,
and offset voltage and gain are interactive. Most importantly, the
input impedance at the negative input is equal to $R_1$ while the input
impedance at the positive input is $R_3 + R_4$. The summing junction
$V_3$ appears as a pseudoground and tracks the positive input $V_4$;
the positive input looks like a simple voltage divider. This configuration works well in circuits such as the one shown in Figure 7.6,
where the input is driven from two amplifier outputs. The amplifier
output looks like a voltage source and is not affected by amplifier
input impedance. This configuration will load a sensor and affect
the resistive temperature-compensation networks around the chip.

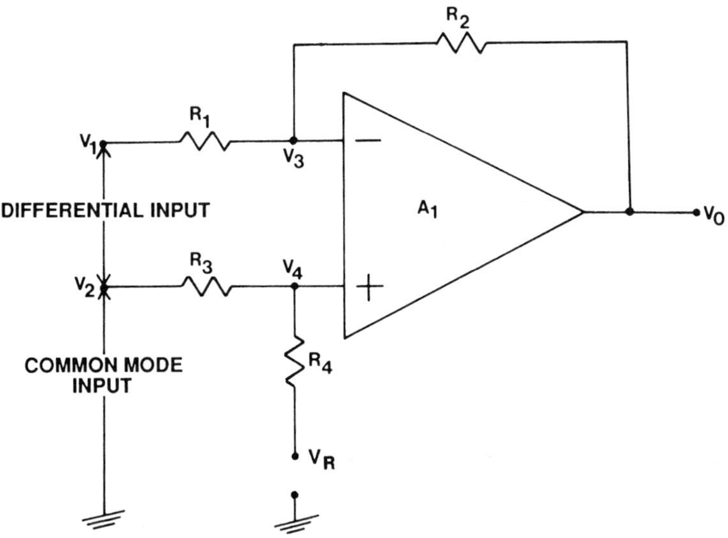

FIGURE 7.6   Difference amplifier.

When a sensor is temperature-compensated, the voltage across
the bridge is changed as a function of temperature, which causes
the sensor output common-mode voltage to change at the same rate.
The amplifier should be sensitive to differential voltage changes
at the chip, but reject any common-mode changes (Figure 7.7).
In this amplifier configuration, the common-mode rejection is determined by the match of the two ratios $R_4/R_3$ and $R_2/R_1$ (Equation 7.5).
If the two ratios are perfectly matched, the output will not change
when the two inputs are shorted together and a changing common-
mode voltage is applied to the inputs. If any one of the four resistors
is changed, the common-mode gain is changed, as well as the differential gain. This can have serious effects over temperature. To
maintain high common-mode rejection, pairs of resistors have to
be changed.

The third problem with this circuit involves the offset voltage.
The differential offset voltage cannot be adjusted at the amplifier,
but must be adjusted before the amplifier or the common-mode bias
voltage can be changed to compensate differential offset. Referring
to Figure 7.6, if the $V_{REF}$ node on $R_4$ is driven from a true voltage
source (0-$\Omega$ output impedance), the output can be increased without
affecting the common-mode rejection of the amplifier. The output
of an operational amplifier can be used to drive this point. The
low-impedance amplifier output will isolate $R_4$ from the rest of the
circuit. The closed-loop output impedance of amplifier $A_2$ of Figure 7.7
can be calculated.

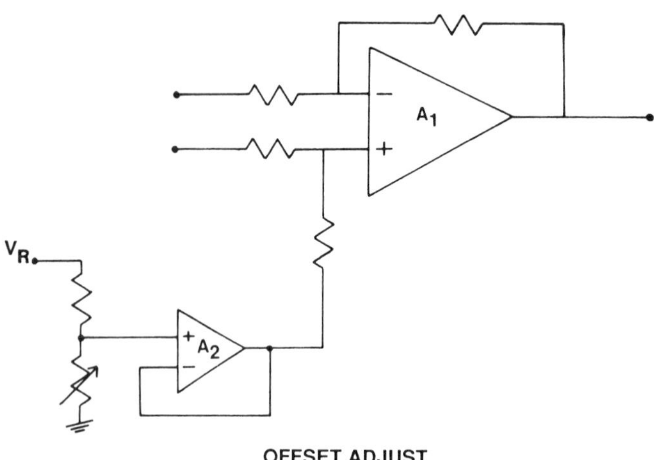



FIGURE 7.7   Offset adjust.


If amplifier $A_2$ has an open-loop output impedance of 200 Ω and
an open-loop gain of 20,000, the effective output impedance is

$$Z_{CL\ OUT} = Z_{OL\ OUT}/A \qquad (7.7)$$

$$Z_{CL\ OUT} = 200/20,000 = 0.01\ \Omega$$

If a single power supply is used, this technique can only compensate
a negative offset voltage, since the reference pin would have to
be driven below ground to compensate a positive voltage. The amplifier has a gain of 1 from the reference pin to the output.

   This amplifier configuration is used in the signal-conditioning
electronics as a fixed differential-gain stage.


## 7.3.4   Three-Amplifier Instrumentation Configuration

The three-amplifier instrumentation amplifier is the most popular
configuration. Figure 7.8 shows the amplifier and resistor configuration. Amplifiers $A_1$ and $A_2$ form a differential input/differential
output amplifier; the inputs are applied to the high-input-impedance
positive inputs. $R_3$ sets the amplifier gain and $A_3$ (difference amplifier) rejects the common-mode voltage and converts the differential
output of the first stage to a single-ended output. The output
of the three amplifiers is as follows.

Amplifier $A_1$ output is

$$V_{OUT} = -(1 + 2R_1/R_G)V_d/2 + V_{cm} \qquad (7.8)$$

Amplifier $A_2$ output is

$$V_{OUT} = (1 + 2R_1/R_G)V_d/2 + V_{cm} \qquad (7.9)$$

Overall output is

$$V_{OUT} = (1 + 2R_1/R_G)V_d \qquad (7.10)$$

$$R_1 = R_2, \text{ and } R_3 = R_4 = R_5 = R_6$$

The match of resistors $R_3$ through $R_6$ sets the common-mode rejection of the amplifier. This circuit is very versatile and will work with large common-mode voltages at the input. The input amplifier's common-mode voltage range determines the maximum common-mode input voltage limits. Low-noise amplifiers can be used for the input amplifiers when a high signal-to-noise ratio is required.

### 7.3.3 Two-Amplifier Configuration

A gain stage with all the characteristics of an instrumentation amplifier can be designed using two operational amplifiers and five precision resistors (see Figure 7.9).

*Gain Calculations*

Neglecting the error terms of the operational amplifier, such as offset voltages and bias currents, the equation for the overall amplifier is given by

$$V_0 = R_4/R_3[1 + \tfrac{1}{2}(R_2/R_1 + R_3/R_4) + (R_2 + R_3)/R_0]V_d$$
$$+ R_4/R_3(R_3/R_4 - R_2/R_1)V_{cm} \qquad (7.11)$$

The output is given as a common-mode voltage, $V_{cm}$, plus a differential input, $V_d$. The ratio of $R_3/R_4$ is made equal to the ratio $R_2/R_1$, to reject common-mode input voltages. The differential input is then amplified according to

$$V_0 = R_4/R_3(1 + R_3/R_4 + (R_2+R_3)/R_0)V_d \qquad (7.12)$$

$$\text{where } R_3/R_4 = R_2/R_1$$

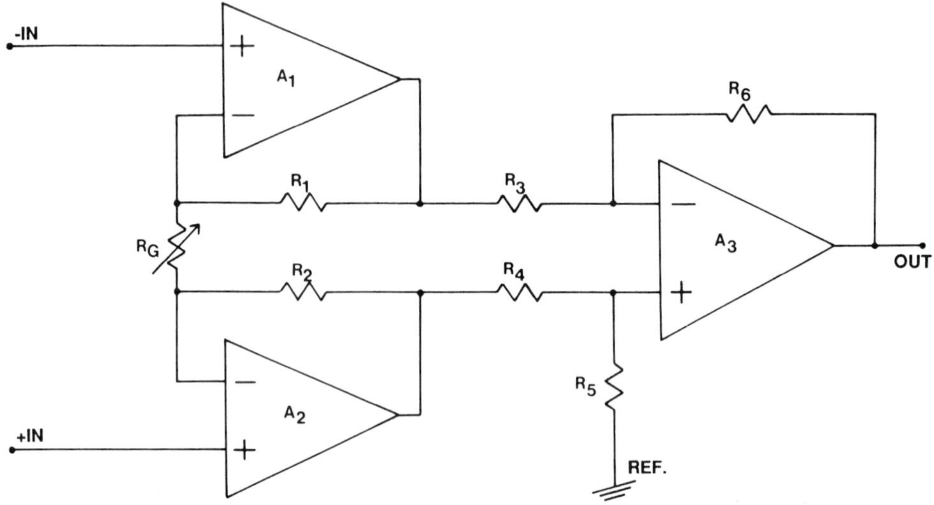


FIGURE 7.8    Three-amplifier configuration.

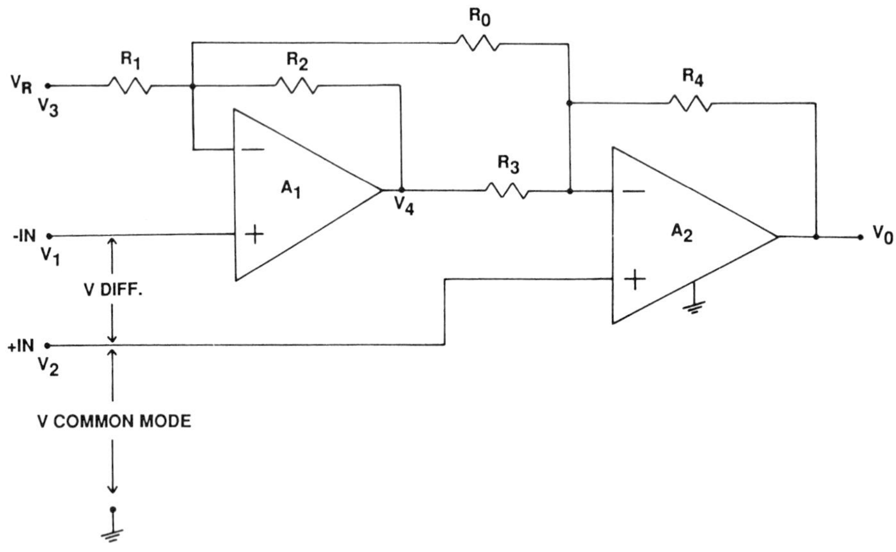


FIGURE 7.9    Two-amplifier configuration.

If the output is to swing close to ground, the ratio of $R_3/R_4$ should be at least 2. This ensures that amplifier $A_1$ will have enough voltage swing to drive the output close to ground. This also allows $A_1$ to be a general-purpose amplifier. The dynamic range of $V_4$ is

$$V_4 = -(1 + R_N/R_0)V_d + 2V_{cm} \qquad (7.13)$$

The problem arises when $V_{cm} = 1/2\ V_{supply}$; amplifier $A_1$ cannot swing to the supply voltage, forcing the output to ground.

If the output does not need to swing close to ground, $R_1$ through $R_4$ can be made equal, simplifying the gain equation:

$$V_0 = 2(1 + R_N/R)V_d \qquad (7.14)$$

where $R_N = R_1 = R_2 = R_3 = R_4$

The $V_{REF}$ input has a gain of 1 to the output. Changing the voltage at $V_{REF}$ changes the output reference voltage (differential input voltage = 0). This can be used to level-shift the output or compensate for temperature errors or to make the circuit ratiometric to the power supply.

### Ratiometric Configuration

Figure 7.10 shows the same basic circuit. If a ratiometric A/D converter is used with a sensor, the sensor and amplifier can be designed to be ratiometric to the supply, allowing the supply voltage to change without affecting the sensor measurement. In the ratiometric circuit (Figure 7.10), the resistor from the amplifier $A_1$ summing junction to the reference voltage is split into two equal resistors exactly twice the value of $R_1$. A voltage divider is created from the power supply to ground, forcing the reference voltage to 1/2 the power supply. If the power-supply voltage is changed, the reference voltage changes proportionally. The amplifier output is given by

$$V_0 = 2(1 + R_1/R_0)V_b + \tfrac{1}{2}V_s \qquad (7.15)$$

### Circuit Limitations

This circuit is useful if the input common-mode voltage is 1/2 the supply voltage.

The common-mode rejection is determined by the ratio of resistors $R_1$ through $R_4$. A small pot can be used in series with $R_1$ to adjust the ratio of $R_2/R_1$ to $R_4/R_3$ to improve the common-mode rejection of the amplifier.

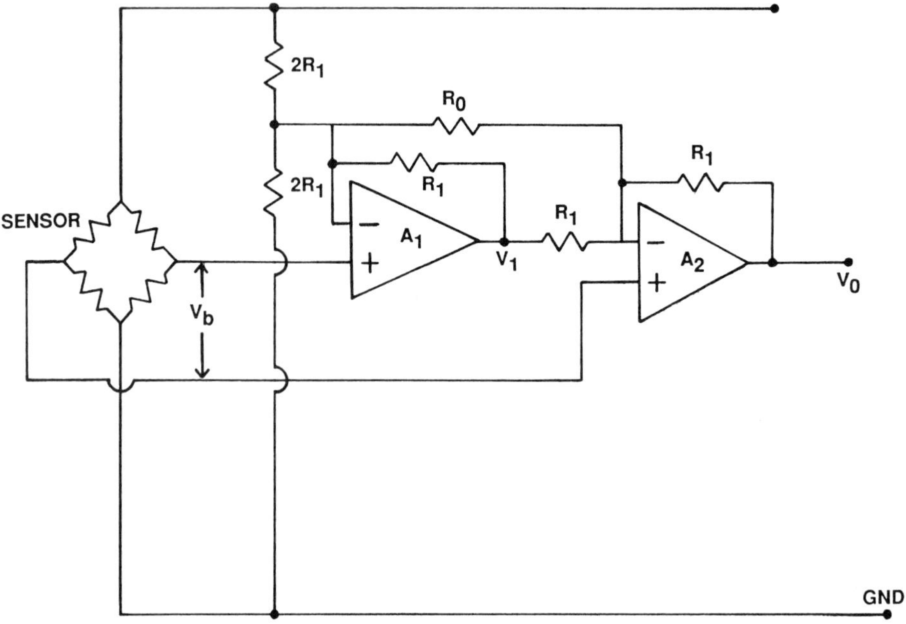


FIGURE 7.10    Ratiometric amplifier.


### 7.3.4  Amplifier Noise

Every electronic component generates some noise voltage: Noise
cannot be eliminated, just minimized. Selecting low-noise components
and being aware of the various noise sources is the most a circuit
designer can do. We have looked at the sensor as a noise source.
The sensor is constructed from passive components and will generate
thermal noise. The amplifier is an active component and is capable
of generating much more noise than is the sensor. If the amplifier
was not selected for low noise, it will probably be the major noise
source in the circuit. The sensor is passive and generates very
little noise. The reference voltage or current source used to power
the bridge and the amplifier used to amplify the sensor output
will generate most of the noise.

*Amplifier Noise Definitions*

The amplifier appears to have two noise sources: a voltage source
that is relatively independent of the sensor impedance and a current
noise that is directly affected by the sensor impedance. The higher

the sensor impedance, the more attention should be paid to the current noise. These noise sources are referred to the *amplifier inputs* and are amplified as part of the input signal.

*Noise Voltage*:  The noise voltage or short-circuit equivalent input noise voltage is the noise that would appear to be generated by the input of the amplifier when the input terminals are shorted together. Figure 7.11a shows a typical curve of the noise spectrum of a general-purpose amplifier. The noise voltage is specified in nanovolts (nV) per root hertz at a specific frequency band. The level of noise voltage is not constant over the frequency band; typically, it increases at lower frequencies, as shown in Figure 7.11a. This increase is called *1/f noise*.

*Noise Current*:  The equivalent open-circuit noise current is measured by operating the amplifier with a known impedance at the input terminals and measuring the noise generated by the noise currents. The equivalent noise current is given in picoamps (pA) or femto-amps (fA) per root hertz at a specific frequency. The noise current also increases slightly at lower frequencies, as shown in Figure 7.11b.

*Noise Specifications*:  You will hear all sorts of terms like 1/f noise, white noise, pink noise, shot noise, popcorn noise, and thermal noise. The one thing that can be counted on is that every component will generate some noise. The total noise voltage generated by the system compared to the signal voltage is the major concern. In most general-purpose applications, noise will not be a problem, but in applications where a small change in pressure is to be measured or a larger pressure is to be measured very accurately, noise will be a major concern. Unfortunately, there is no fixed standard as to how noise is specified. Some manufacturers put a plot of equivalent noise voltage vs. frequency in the data sheet, while others give a plot of noise voltage and noise current vs. frequency. To get a comparison, we will now look at the noise-voltage curves of an amplifier designed for low noise, a general-purpose amplifier, and an FET in-stage amplifier. Figure 7.12a is the noise-voltage plot of a low-noise amplifier, Figure 7.11a is the noise-voltage plot of a general-purpose amplifier, and Figure 12b is the noise-voltage plot of an FET input amplifier.

### Calculating Total Noise

We can generate a plot of total noise for various values of $R_b$ if noise voltage and noise current are known versus frequency. The noise voltage is calculated by taking the square root of the sum of the squares, as shown in Equation 7.16:

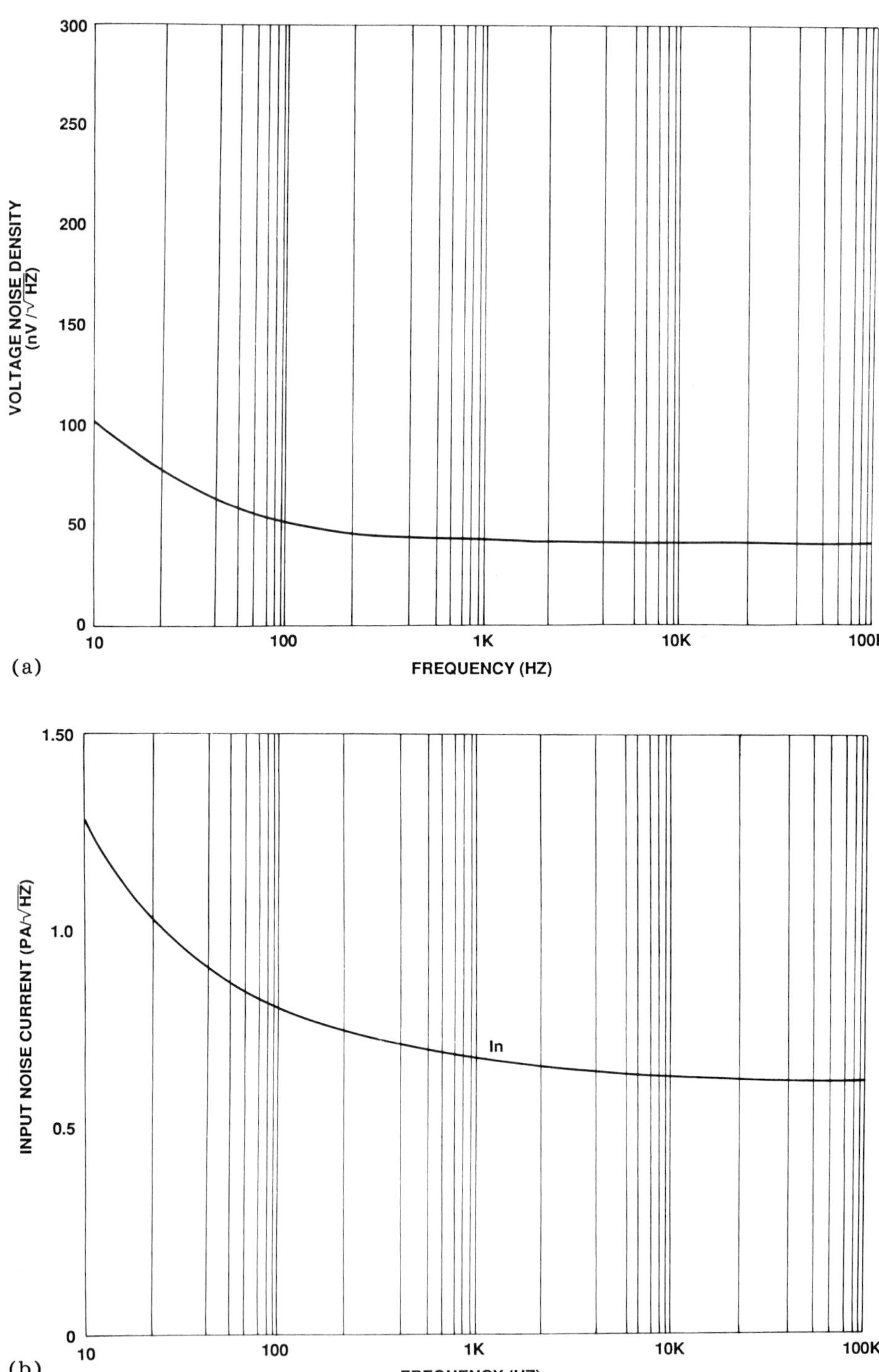


FIGURE 7.11  (a) General-purpose amplifier—noise voltage.
(b) General-purpose noise current.

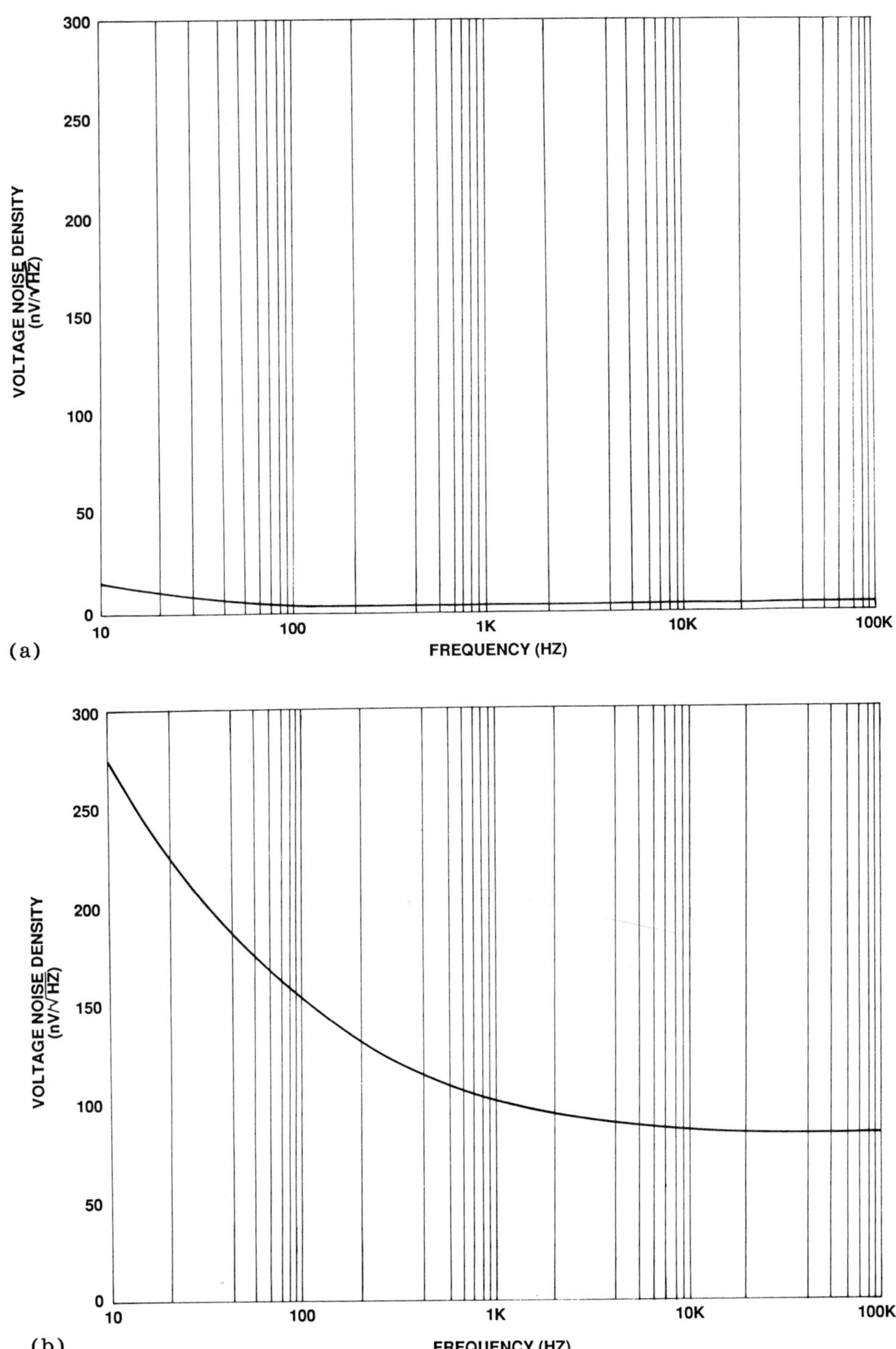


FIGURE 7.12   (a) Low-noise amplifier—noise voltage. (b) FET input amplifier—noise voltage.

$$e_N = (e_n^2 + e_R^2 + (i_n \times R_b)^2)^{1/2} \qquad (7.16)$$

A graph of noise voltage is shown in Figure 7.13. To make the plot, the thermal noise must be calculated from Equation 7.13:

$$e_R = 4ktRB^{1/2} \qquad (7.17)$$

where $e_R$ = noise voltage

$\qquad$ k = Boltzmann's constant ($1.38 \times 10^{-23}$ joule/K)

$\qquad$ t = absolute temperature (degrees Kelvin)

$\qquad$ R = bridge resistance

$\qquad$ B = frequency bandwidth

The amplifier noise voltage and noise current are taken from the amplifier data sheet, and the bridge impedance is given on the sensor data sheet. Figure 7.13 shows graphs for 100-$\Omega$, 1-k$\Omega$, 10-k$\Omega$, and 100-k$\Omega$ bridge impedances for one amplifier. The difference

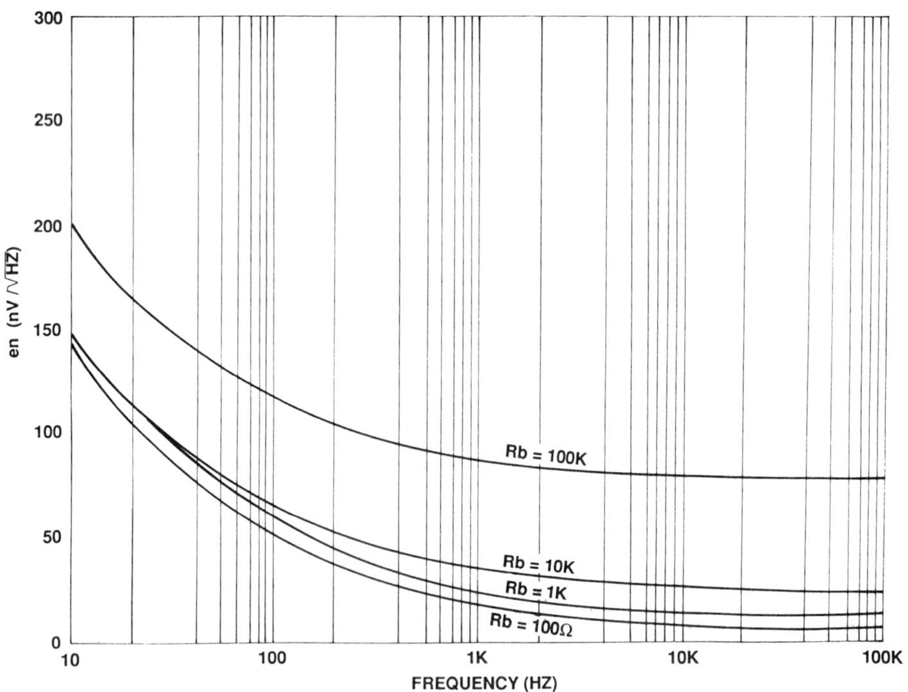


FIGURE 7.13  Total noise voltage.

in the graphs is due to the noise voltage generated by the current noise times the source impedance. The FET input amplifier has more voltage noise generated at the input, but much less current noise. Due to the low bias current, FET amplifiers are used when very high source impedances are used.


## 7.4  SENSOR EXCITATION

The two means of providing excitation for pressure sensors are constant current and constant voltage. The characteristics of constant current and constant voltage sources are as follows:

1.  Constant Current Source
    a. High output impedance
    b. Fixed output current
    c. Supplies a constant current over a wide voltage range
2.  Constant Voltage Source
    a. Low output impedance
    b. Fixed output voltage
    c. Supplies a fixed output voltage over a wide output current
       range

The excitation used with a particular sensor will depend on the technique used to temperature-compensate span. Sensor span is defined as gage factor in mV/V/psi × bridge voltage × the applied pressure in psi:

$$S = GF \times V_b \times psi \qquad\qquad (7.18)$$

where S = span

GF = gage factor

$V_b$ = bridge voltage

psi = applied pressure

The sensor sensitivity (gage factor) has a negative temperature coefficient, generally in the range of −2000 ppm/°C. From Equation 7.18, if $V_b$ is forced to have a positive temperature coefficient equal in magnitude to the gage-factor temperature coefficient, the span will be temperature-compensated. This is the basic technique used to temperature-compensate piezoresistive pressure sensors independent of excitation.

The temperature coefficient of the bridge is the key parameter when deciding on a technique for compensating span. If the temperature coefficient of the bridge is larger than the gage-factor temperature coefficient, the sensor can use constant voltage or constant current excitation. If the temperature coefficient of the bridge is less than the gage-factor temperature coefficient, the sensor must use constant voltage and series thermistors or a temperature-dependent current source to compensate span.

Sensors that use constant current excitation have the advantage that temperature is measured by the sensor chip, thus reducing temperature gradients. If a sensor is used to measure a medium that may be at extreme temperatures, a sensor that has multiple temperature sensors—such as thermistors or diodes—can have large temperature gradients from the chip to ambient temperature. A good example is a refrigeration application, where a sensor is used to measure the freon pressure; the freon will be much cooler than ambient temperature. Next, we will discuss how to design constant voltage and constant current sources.

### 7.4.1  Constant Voltage Source

Any voltage regulator meets the requirements of a constant voltage source. It has low output impedance and will supply any current necessary to keep the load at a constant voltage. A number of voltage sources can be used to power the sensor: a standard voltage regulator, a zener diode, or a buffered reference.

*Standard Voltage Regulators*

The problems encountered with most voltage regulators are that just a few voltages are available, and most regulators have a fairly large temperature coefficient. Since piezoresistive sensors are ratiometric to the supply voltage, any change in the supply voltage will cause a proportional change in the sensor sensitivity. Here again, the accuracy needed for the application will determine which type of voltage source should be used. An example of how a voltage regulator could be used is shown in Figure 7.14. The 15-V system power is regulated down to 8 V by an on-card regulator the 8 volts is used to power just the sensor and amplifier bias networks. The regulator also isolates the sensor from the noise and voltage spikes that may be on the system power. This approach should be considered for 2 percent to 5 percent total error systems.

*Zener Diodes*

A standard zener diode can be used as a shunt regulator, as shown in Figure 7.15. Like the voltage regulators, zeners are available in limited voltages and may have a relatively large temperature

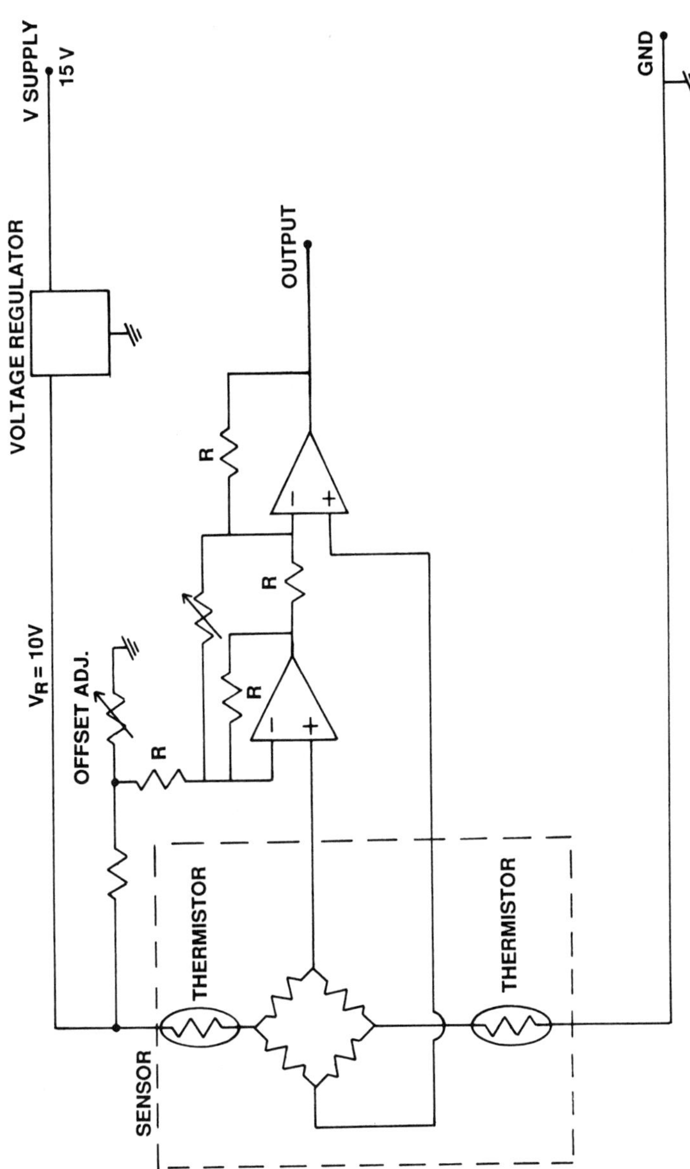


FIGURE 7.14   Voltage regulator used as voltage source.

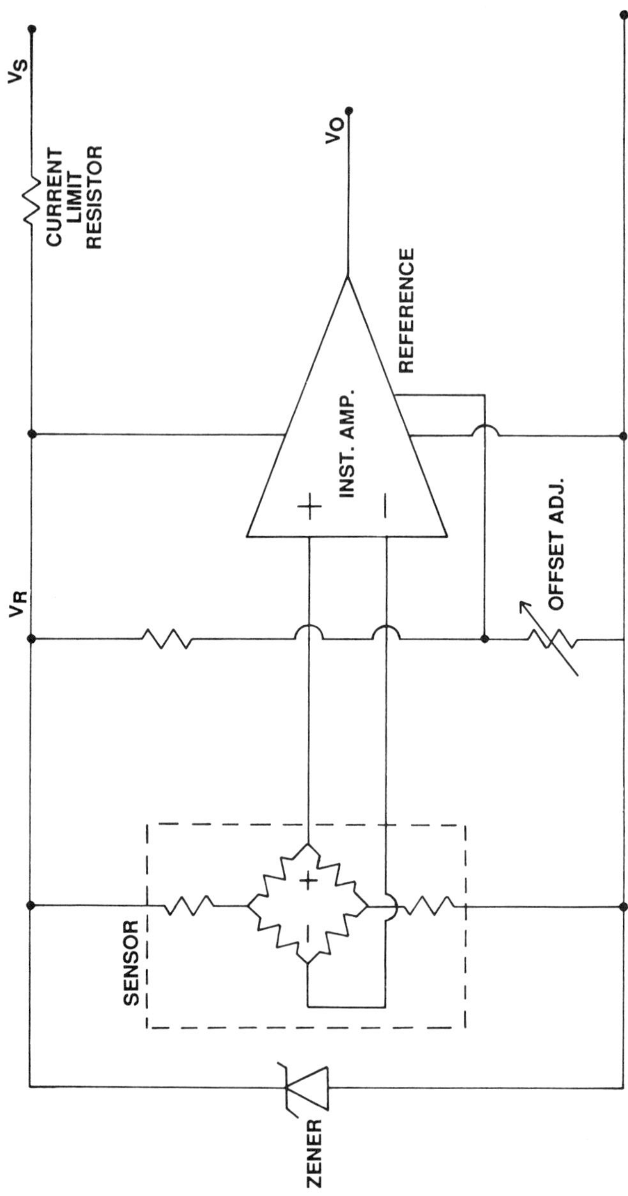


FIGURE 7.15  Zener diode voltage source.

coefficient. Zener diodes are readily available in voltages over 6 V but are not easy to find in voltages of 5 V or less.

### Buffered Reference

In the past few years, a number of semiconductor manufacturers have introduced voltage reference integrated circuits. These band-gap devices have all the important parameters needed for a voltage reference:

1. Low noise
2. Excellent temperature stability
3. Long-term stability
4. Low dynamic impedance
5. Low DC impedance
6. Operate over a wide range of current

*Noise*:  These references have very low noise—in the range of 10 to 20 µV peak to peak. Low noise is necessary when designing very-low-pressure systems.

*Temperature Stability*:  Temperature stability is in the range of 10 to 25 ppm/°C.

*Long-Term Stability*:  This is a very important parameter when designing precision pressure transducers. Perfect linearity and temperature characteristics do no good if the device drifts with time.

*Low Dynamic Impedance*:  The reference should filter out any spikes on the power supply; the lower the dynamic impedance, the better the regulation.

*Low DC Impedance*:  If the supply voltage changes, the current through the reference will change proportionally. If the reference has a high impedance, the voltage will change by $\Delta I \times Z$.

*Operate Over Wide Range of Currents*:  Band-gap references will operate from 10 µA to 10 mA. Operating down around 10 µA is important if you are designing a low-power system, and the wide operating current range allows the reference to operate from a wide range of supply voltages.

Band-gap references are available at 1.2-, 2.5-, and 5-V operating voltages. Using a 1.2-V reference and amplifying the output with an operational amplifier allows the designer to have any voltage from 1.2 V to the supply voltage. Figure 7.16 shows a band-gap reference and buffer amplifier. The amplifier output will look like a true voltage source with low output impedance and will supply up to 10 mA to the load. $Z_1$ is the band-gap reference IC (1.2 V).

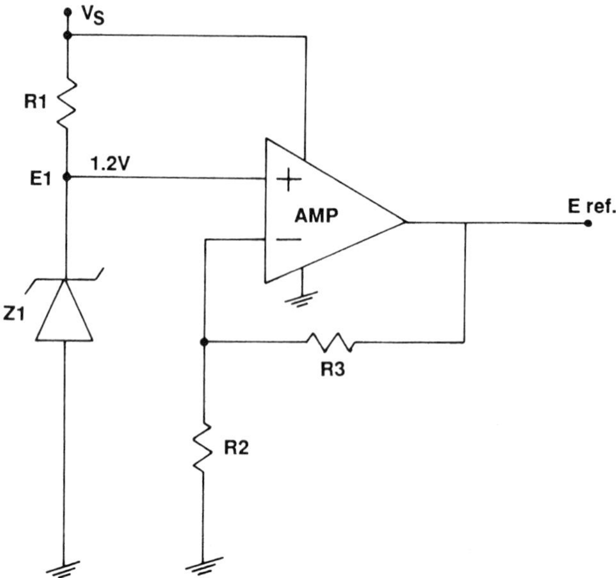


FIGURE 7.16   Buffered band-gap reference.


$R_1$ sets the reference device current over the operating voltage range. To calculate the value of $R_1$:

$$R_1 = V_S - 1.2/I_{REF} \tag{7.19}$$

*Example*:   The supply is 15 V, $I_{REF}$ is 1 mA, and the device must operate over a voltage range of 8 to 24 V, with normal operation at 15 V. The band-gap reference will operate over a range of current from 10 μA to 10 mA.

$$R_1 = 15 - 1.2/1 \text{ mA}$$

$$R_1 = 13.8 \text{ k}\Omega$$

At 8 V:

$$I_{REF} = 8 - 1.2/13.8 \text{ k}\Omega$$

$$I_{REF} = 493 \text{ μA}$$

At 24 V:

$$I_{REF} = 24 - 1.2/13.8 \ k\Omega$$

$$I_{REF} = 1.65 \ mA$$

A value of 13.8 k$\Omega$ will set the current to 1 mA at 15 V and keep the current in the reference operating range from 8 to 24 V.

The output voltage is set by the ratio of resistors $R_2$ and $R_3$:

$$E_{REF} = e_1 \ (1 + R_3/R_2) \tag{7.20}$$

Any voltage between 1.2 V and $V_S$ - 2 can be set by adjusting the ratio of the two feedback resistors.

*Example*: An 8-V reference is needed to power a sensor. Using the circuit in Figure 7.16 and Equation 7.20, $Z_1$ is a 1.2-V band-gap reference and $R_2$ is 10 k$\Omega$. $R_3$ is calculated as follows:

$$R_3 = E_{REF} - e_1/(e_1/R_2) \tag{7.21}$$

$$R_3 = 56.67 \ k\Omega$$

To make an 8-V reference, $R_2$ = 10 k$\Omega$ 1 percent and $R_3$ = 56.6 k$\Omega$ 1 percent resistors.


## 7.4.2  Constant Current Source

Constant current sources, as previously defined, supply a constant current over a voltage range. A constant current source requires a reference voltage or current to set the current and an amplifier to supply the current to the load. There are a number of options for designing a circuit using a current-excited sensor: using FET current regulators, designing the sensor in the feedback loop of an operational amplifier, or using an operational amplifier as a voltage-to-current converter.

### Field Effect Transistor

If the gate on an FET is connected to the source, the device will regulate the current at $I_{dss}$. The FET will maintain this current independent of supply voltage (see Figure 7.17a). If a small resistor is placed in the source of the FET, the current can be adjusted (see Figure 7.17b).

FETs can be purchased in fixed current ranges and will operate from 1 to 40 V. The advantage of the FET current source is that it is one component, it needs only about 1 V to operate, and it is temperature compensated internally. The disadvantage of the FET

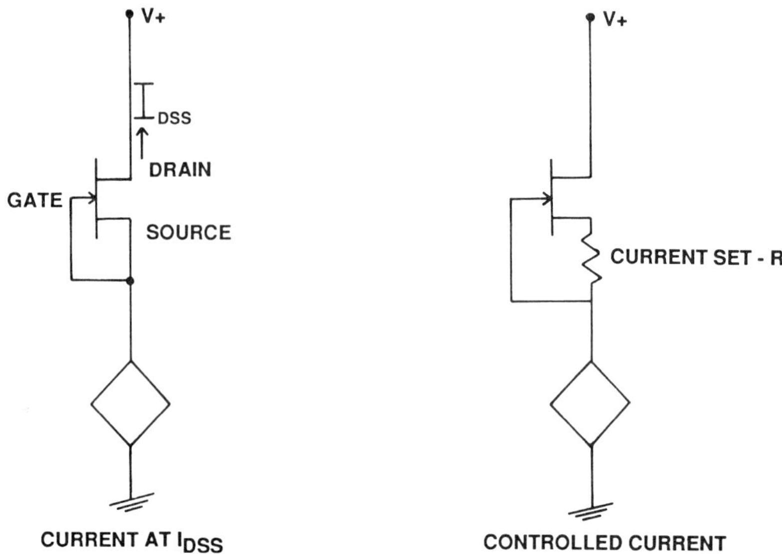


FIGURE 7.17   FET current source.

is that the current can vary ±5 percent, and the temperature coefficient is not constant for all current ranges. The FET is a good
choice for a low- to moderate-accuracy application or an application
that has a low supply voltage.

*Feedback Loop*

The most popular way to design a constant current source is to
put the bridge in the feedback loop of an operational amplifier (see
Figure 7.18).  The bridge current is

$$I_b = e_1 / R_1 \qquad\qquad (7.22)$$

In this design, a band-gap reference is used as a reference voltage
to set the current, and resistor $R_1$ sets the current. $R_1$ should
be a precision resistor with a low temperature coefficient. This
is a fixed current source; the same basic current source can be
made ratiometric to the power supply by referencing it to a voltage
divider, as shown in Figure 7.19.   $R_1$ sets the bridge current,
but $e_1$ is ratiometric to $V_S$.

*Example*:   A fixed current source to supply 1.5 mA to a 5-k$\Omega$ bridge:

$Z_1$ is 1.2 V, $V_S$ is 12 V

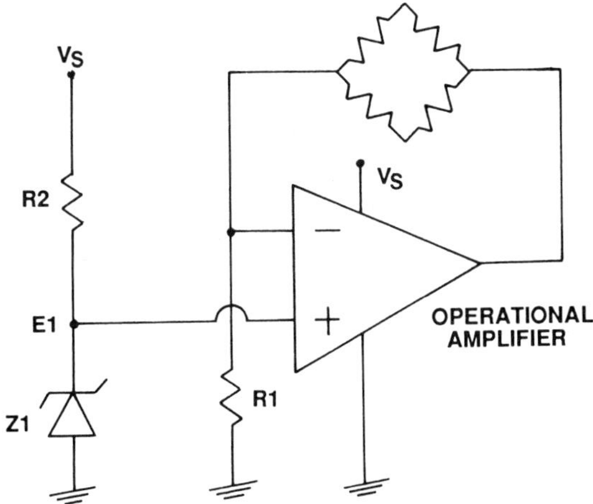



FIGURE 7.18  Fixed constant-current source.

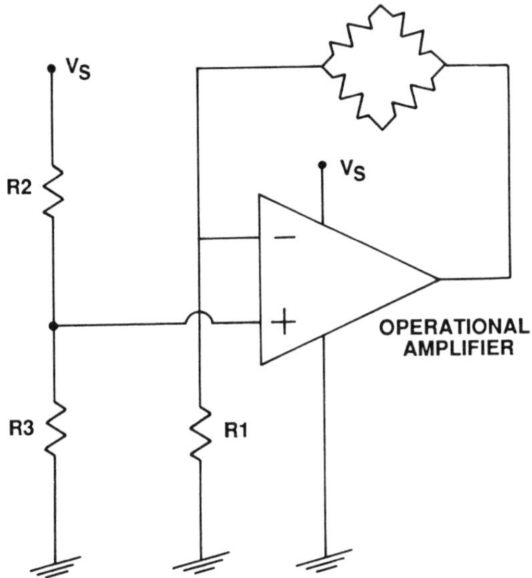



FIGURE 7.19  Ratiometric current source.

$$e_1 = 1.2 \text{ V}$$

$$R_1 = e_1/I_b = 1.2/1.5 \times 10^{-3} = 800 \ \Omega$$

Care should be taken that the amplifier will not saturate at the minimum supply voltage. The amplifier output will be $e_1 + V_b$ (1.2 + 7.5 = 8.7 V at 25°C), and the bridge voltage has a positive temperature coefficient of 2100 ppm/°C. The amplifier will saturate at $V_S$ = -1.5 V. The bridge voltage at the maximum operating temperature should be used to calculate the voltage compliance. In this configuration, the bridge voltage is fixed by the amplifier and reference voltage.

### Voltage-to-Current Converter

If the bridge needs to be floating, a third type of current source can be used (Figure 7.20). In this configuration, the load is driven

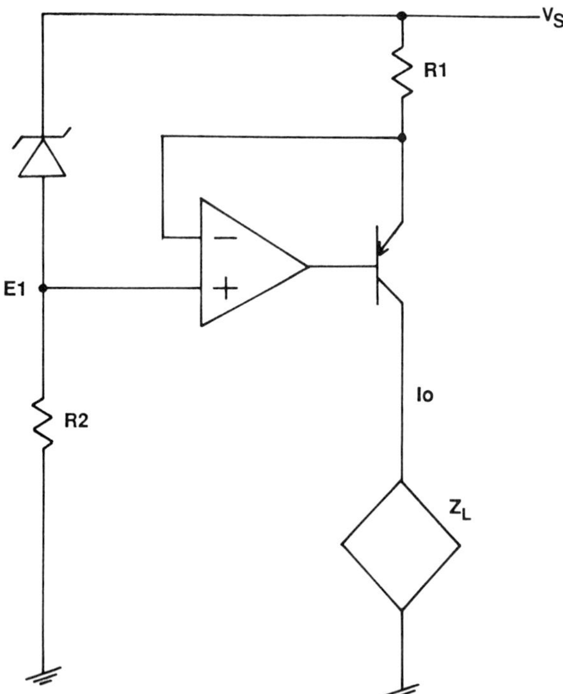


FIGURE 7.20   Voltage-to-current converter (transistor).

by the collector of a transistor, and the operational amplifier is used as a voltage-to-current converter. The load is free to move up and down from ground without affecting the bridge voltage. This configuration is closer to the ideal current source, because the transistor collector presents a high output impedance to the load:

$$I_0 = \alpha(V_S - e_1/R_1) \tag{7.23}$$

$$Z_0 = R_C \;||\; C_C \tag{7.24}$$

This configuration can also be made ratiometric by replacing reference $Z_1$ with a precision resistor. From Equation 7.23, the alpha of the transistor represents an error term, and not all the current set by $R_1$ gets to the load. The transistor has a small base current. To minimize this effect, the transistor should have as high a $\beta$ as possible. The same basic circuit can be designed using FET transistors, thus eliminating the $\alpha$ term (see Figure 7.21). The output current of this configuration is given by

$$I_0 = V_S - e_1/R_1 \tag{7.25}$$

The output impedance will be of the order:

$$Z_0 \simeq 10^{12} \; \Omega \;||\; 3pf$$

### 7.4.3  Design Example

We have discussed the individual components necessary to signal-condition a pressure sensor. Next, we will design the complete circuit.

A pressure sensor is needed to measure a pressure range of 0 to ±2.5 psig and interface to a computer through a 10-bit A/D converter. The A/D converter has a 0- to 5-V input range, and the system should have a static accuracy of 0.5 percent (25°C accuracy), a dynamic accuracy (over temperature) of ±2.5 percent, and a temperature range of 0 to 70°C. Twelve volts is available for power.

First, a sensor is selected with the following specifications:

Pressure range: 0-5 psig
Excitation: 1.5 mA DC
Offset: ±2 mV
Span: 75-125 mV (at 1.5 mA)
Bridge resistance: 5000 $\Omega$

Linearity: 0.2% FS
Offset temperature error: ±1% (0-70°C)
Span temperature error: ±1% (0-70°C)
Repeatability: ±0.1% FS
Printed circuit board mountable

From the specification, the static accuracy is ±0.3 percent (0.2 percent linearity + 0.1 percent repeatability) and the dynamic accuracy is ±2 percent (±1 percent offset + 1 percent span). The next step is to design a current source for the sensor. The sensor is specified at 1.5 mA and has a 5-kΩ impedance. At room temperature, the voltage across the sensor will be 7.5 V. The 5-kΩ impedance is a worst-case specification. The bridge is paralleled with a resistor to adjust the TCR to be equal to TC-gage factor. Paralleling the bridge will decrease the total impedance; the size of the parallel

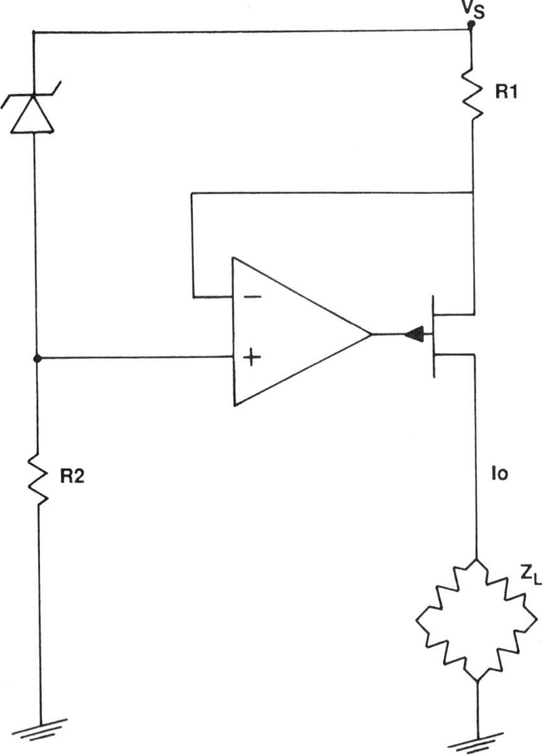



FIGURE 7.21  Voltage-to-current converter (FET).

resistor will depend on how close TCR and TC-GF are initially. Assuming worst-case impedance at 70°C, the bridge voltage will be 8.21 V (2100 ppm × 7.5 V × 45°C + 7.5 V). Before we design the current source, an operational amplifier has to be selected. The A/D converter needs a 0- to 5-V input, so an amplifier that will swing to ground is needed.

Next, we select an operational amplifier for the current source and the instrumentation amplifier. For this application, we will select a quad general-purpose amplifier with the following specifications:

Input offset voltage: ±500 µV
Input offset current: 3.0 nA
Input bias current: 30 nA
Open-loop gain: 2 V/µV
Input voltage range: $V_E$ - 1.2 V to -0.3 V
Output voltage swing (low): 150 mV (2K load)
Output voltage swing (high): $V_E$ - 0.6 V
Supply current: 4 mA (4 amplifiers)

There are typical specifications for a medium-priced amplifier.

The current source and amplifier are shown in Figure 22. Amplifier $A_1$ is the current source, $A_2$ and $A_3$ make up the instrumentation amplifier, and $A_4$ is used as an offset adjust. The values for $R_1$ and $R_2$ are calculated:

$$R_1 = 12 - 1.23/1.0 \times 10^{-3}$$

$$R_1 = 10.77 \text{ k}\Omega, \text{ or } 10 \text{ k}\Omega$$

Resistor $R_2$ and band-gap reference $Z_1$ set the current through the bridge, and $Z_1$ provides a stable 1.2-V reference to the positive input of amplifier $A_1$. Resistor $R_1$ sets the current to the summing junction:

$$I_B = 1.23/R_1 \tag{7.26}$$

$$R_1 = 1.23/1.5 \times 10^{-3}$$

$$R_1 = 820 \ \Omega$$

Thus, 820 $\Omega$ sets the current through the bridge to 1.5 mA.

The next step is to check the voltage compliance of the amplifier. We assume some typical values for compensation resistors $R_3$, $R_4$, and $R_5$—$R_3$ = 12 k$\Omega$, $R_4$ = 12 $\Omega$, and $R_5$ = 260 k$\Omega$. $R_3$ is the span-temperature-compensating resistor, which adjusts the TCR of the bridge to be equal to TC-gage factor and, when installed, reduces the impedance of the total bridge. In this case, the bridge impedance will be

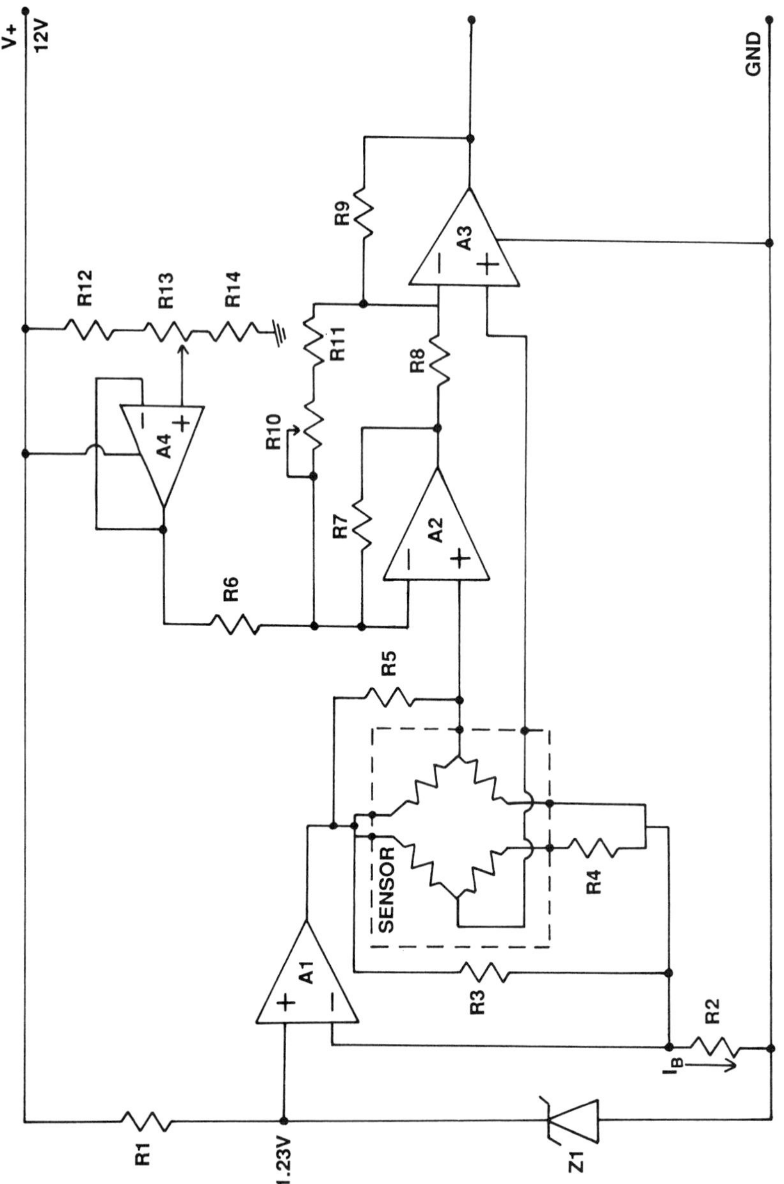

FIGURE 7.22  ±2.5-psig transducer.

$$Z_B = R_B \,||\, R_3 \qquad (7.27)$$

$$1/Z_B = 1/5 \text{ k}\Omega + 1/12 \text{ k}\Omega$$

$$Z_B = 3.529 \text{ k}\Omega$$

This sensor will have a voltage compliance of

$$V_B = V_{REF} + I_B \times R_B \qquad (7.28)$$

$$V_B = 1.23 + 1.5 \times 10^{-3} \times 3529 \ \Omega = 6.523 \text{ V}$$

The output of amplifier $A_1$ will swing to approximately 10.5 V; this is the calculation for a typical sensor. The worst-case calculation would be if $R_3 = \infty$, the 70°C bridge impedance would be

$$Z_B = ((5K \times 2 \times 10^{-3}) \times 45) + 5K$$

$$Z_B = 5.45 \text{ k}\Omega$$

and the worst-case voltage compliance (using Equation 7.28) is

$$V_B = 1.23 + 1.5 \times 10^{-3} \times 5.45 \times 10^3$$

$$V_B = 9.40 \text{ V}$$

Voltage compliance will not be a problem.

Resistor $R_4$ is the offset-adjust resistor, which sets the output to within ±2 mV at 0 psig. Compensation resistor $R_5$ is the offset-TC compensation resistor.

The next step is to set up the instrumentation amplifier. With no pressure applied, the output will be 2.5 V. The amplifier will have a gain of

$$A_V = \Delta V_{OUT}/\Delta V_{SENSOR} \qquad (7.29)$$

$$A_V = 2.5/.05$$

$$A_V = 50$$

Now that we know the gain needed, we can set up amplifier $A_4$ (offset-adjust amplifier). The sensor will have a worst-case offset of ±2 mV with 0 psig applied. Sensor worst-case offset plus amplifier offsets will be ±3 mV at the input or ±150 mV at the amplifier output. Amplifier $A_4$ has a gain of 1 to the transducer output, and we will set up $R_{13}$ to have a range of 2.5 V ± 2 V, or 2.3 to 2.7 V minimum. Potentiometers are available in 100 $\Omega$, 200 $\Omega$, 500 $\Omega$, and 1 k$\Omega$. The value of the potentiometer should be chosen first, due to the limited choices of values. A 20-turn 500-$\Omega$ pot will be used for $R_{13}$. $R_{14}$ = 2.4 k$\Omega$, and $R_{12}$ = 10 k$\Omega$. These values give an adjustment range of 2.232 to 3.697 V at the output of amplifier $A_4$.

The last thing to do is to set up the instrumentation amplifier. Amplifiers $A_2$ and $A_3$ will be used to make the instrumentation amplifier. The four feedback resistors are made equal: $R_6$ = $R_7$ = $R_8$ = $R_9$ = 100 k$\Omega$. The gain equation for this amplifier is

$$V_O = 2(1 + (R_6/R_O))V_d \qquad (7.30)$$

where $R_O = R_{10} + R_{11}$

$$A_V = V_O/V_d = 2(1 + R_6/R_O)$$

and

$$R_O = 2R_6/A_V - 2$$
$$R_O = 200K/50 - 2 = 4.167K$$

The center value of $R_O$ is 4.17K, and the sensor can have a sensitivity of 75 to 125 mV. The gain of the amplifier needs to be adjustable from 40 to 66.7. A 4.2K resistor and a 2K pot will give a range of 40 to 66.

To adjust the transducer, 0 psig is applied to the sensor and $R_{13}$ is adjusted for 2.5 V at the output. A pressure of 2.5 psig is applied to the sensor, and $R_{10}$ is adjusted for 5.0 V at the output. When -2.5 psig is applied to the sensor, the output will try to go to ground, but will saturate around 50 mV. This design will meet the stated requirements.

# 8

# Sensor Selection

## 8.1 INTRODUCTION

The number of new pressure-sensor applications is increasing due
to the decrease in the price of sensors. Many manufacturers could
improve the performance of a piece of equipment or process by
using a pressure sensor to measure barometric pressure or altitude,
but have not done so because of the high price of sensors. With
the introduction of batch-processed sensors, sensor price is decreasing to the point that it is practical to consider designing in these
improvements for many applications. For example, barometric pressure
can have a direct effect on the operation of a precision laser system
and on many pieces of medical diagnostic equipment. If the system
designer is not familiar with pressure-transducer terminology and
specifications, selecting the right sensor for an application can
be a major task. The first thing the designer needs to do is define
the sensor needed. To do this, the following questions need to
be answered:

1. Application (type of measurement)
   a. Absolute
   b. Gage
   c. Differential
   d. Sealed reference
2. Accuracy requirements
   a. Static accuracy (25°C)
      (1) 0.1% to 1% FS
      (2) 1% to 2% FS
      (3) 2% to 10% FS

      b. Accuracy over temperature
        (1) 0.1% to 1% FS
        (2) 1% to 2% FS
        (3) 2% to 10% FS
3. Electrical
    a. Output voltage
    b. Excitation
    c. Power requirements
4. Operation
    a. Pneumatic or hydraulic
    b. Ruggedness
    c. Media compatibility
5. Temperature requirements
    a. Temperature specifications
    b. Electrical temperature considerations
    c. Mechanical temperature considerations
6. Speed—dynamic measurement
    a. Transients
    b. Acoustic
7. Installation
    a. Location and position
    b. Pipe size and thread
    c. Pressure seals
    d. Access
8. New design or redesign
    a. Redesign: Does the device need to be fit, form and function, to an existing sensor?
    b. New design
      (1) Maximum package size and weight
      (2) Sensor compatibility with system plumbing
9. Costs
    a. Purchase price
    b. Calibration and compensation labor
    c. Capital equipment required
    d. Maintenance

The check list can be broken down into four categories: application (items 1, 4, 6); specifications (items 2, 3, 5); mechanical (items 7, 8); and price (item 9). Answering these questions will define the sensor required for the application.

## 8.2  APPLICATION

What type of measurement is required?

### 8.2.1   Absolute, Gage, or Differential Sensor

*Absolute*

Any measurement that is referenced to a fixed reference pressure
(generally a hard vacuum) will use an absolute sensor. The more
common absolute applications include altimeters, barometers, and
vacuum measurements.

In some applications, an absolute sensor may be used for con-
venience. Since the sensor has the reference pressure sealed inside
the chip, it requires only one pressure input. In applications such
as depth measurements, the sensor can be submerged in liquid
without the need of a vent (reference pressure) tube to the surface.
The disadvantage of this approach is that the surface is at barometric
pressure, which is also measured by the sensor. To get an accurate
depth measurement, the barometric pressure must be subtracted
from the sensor output.

*Gage*

Any measurement that is referenced to barometric pressure will
use a gage sensor. The sensor will measure positive and negative
pressures (pressures greater than and less than barometric pressure).
Gage sensors are useful in equipment that may be taken from one
altitude to another, because the sensor automatically compensates
for any change in barometric pressure (barometric pressure is the
reference pressure). Airplanes and automobiles can measure airspeed
and engine vacuum at sea level and at 10,000 feet with the same
accuracy.

*Differential*

A measurement where one pressure is measured relative to a second
pressure will use a differential sensor. Since gage measurement
is actually a special type of differential measurement, the same
basic sensor is used for both sensors. However, a second pressure
port is added to the differential sensor. Differential sensors are
used to measure flow.

### 8.2.2   Operation

The type of operation the pressure sensor has to perform will dictate
some of the mechanical and electrical parameters.

*Hydraulic or Pneumatic*

Hydraulic applications are generally higher pressure, have threaded-
pipe pressure ports, and require mechanical considerations such as

higher overpressure (possible water hammer), media compatibility, and a mechanically stronger package.

Pneumatic applications are generally lower pressure and, in some applications, the sensor may be mounted on printed circuit boards. These devices are less concerned with overpressure and rugged packaging, but may require small size, low leakage, and light weight. Pneumatic sensors may have tubes or barbed pressure ports that are used with flexible tubing for low-pressure applications. If the medium being measured is clean, dry air, media compatibility is not a concern. High humidity that can condense in the system is a potential problem.

*Ruggedness*

If the sensor is mounted on a printed circuit board inside a cabinet, just about any type of sensor package will work. Off-road vehicle owners may use an axe and a fire hose to clean the mud out of the running gear, or use the sensor as a step as they climb around the vehicle.

*Media Compatibility*

Media compatibility is one of the hardest parameters to pin down. Obviously, environments such as salt water and harsh acids need to be isolated from the sensor. At the other extreme, clean, dry gases can be used with any sensor. The problem applications are those that may work 90% of the time; but if the conditions are just right, the sensor could fail. A compressed-air system that has filters and a dryer installed should produce clean, dry air; but if the dryer is turned off or fails and water condenses in the lines, an unprotected sensor could fail.

### 8.2.3  Speed

In some applications, very slow sensors are needed. In systems where large pressure spikes and pulses from motors and valves are present, the low-pass filtering of a slow sensor will average the pressure reading. In these applications, a fast transducer wouldn't be readable—the reading would jump around following the pressure variations.

In other applications, such as engine diagnostics and measuring heart rate, fast sensors are required to get an idea of the shapes of the pressure pulses. Fast sensors can also be used in audio equipment as musical instrument pickups. The sensor can be mounted inside the instrument and will give a flat frequency response from

DC to >20 kHz. Mounting the sensor inside the instrument eliminates external noises and valve noises of the instrument itself.


## 8.3  SPECIFICATIONS

Sensor specifications may or may not be available at the beginning of a project. If the application has never been done before, some experimental tests may have to be performed. Parameters such as pressure range, accuracy, and speed may have to be determined by building a prototype and testing it. A little testing early in the project can save a lot of headaches later on.


### 8.3.1  System Accuracy

The application will dictate how accurate the pressure-measuring device needs to be. Examples of applications that require accurate sensors are

- Altimeters
- Barometers
- Blood-pressure monitors
- Process-control applications

These applications require 1 percent or better accuracy from the transducer. The first three applications have to measure pressure to function; therefore, selecting the right pressure sensor is critical to the success of the product. These are also products that are price sensitive—the sensor is probably the most expensive component in the system, and its price will have a direct effect on the selling price of the unit. A process-control application, on the other hand, is not as price sensitive; the sensor is considered an overhead expense.

Applications that require less accuracy may include

- Tire pressure
- Oil pressure
- Water pressure
- Air pressure

These applications can use sensors with 5 percent accuracy and are used to monitor the condition of the system. An oil-pressure gage gives an indication of the condition of the lubricating system— the condition of the pump, major leaks, or amount of oil. A minimum

working oil pressure is established, and the meter is used to give
a visual indication if the pressure is above or below the limit. The
engine will function whether the pressure sensor is installed or not.


### 8.3.2  Error Budget

Unfortunately, when most system designers start working on a
new system, they start with the worst-case error budget at the
output and work back to the input. After subtracting the error
of each component from the total error, the remainder is the worst-
case error allocated for the input devices. This can result in a
very tight specification on the input sensor, which increases the
purchase price significantly. In some cases, a review of the components used in the system can lower the overall system component
costs without decreasing the system accuracy.


### 8.3.3  Accuracy Trade-off

The error budget is the one thing the design engineer has control
over. His or her job is to select components for maximum accuracy
and reliability at minimum expense. Assume that the basic pressure-
measuring system shown in Figure 8.1 has a total accuracy require-
ment of ±3 percent over a temperature range of 0°C to 70°C. The
error budget for the entire system is ±3 percent. The next step
is to determine the maximum error for each function. The input
device (sensor), signal-conditioning stage, and signal converter
(analog-to-digital converter) device will each need to be considered.
The amplifier must be adjusted for offset and sensitivity and have
a temperature coefficient of 100 ppm/°C, which is +0.4 percent,
-0.2 percent error over temperature.

*A/D Converter*:  If an 8-bit analog-to-digital converter is used,
   the resolution is 1/256, or 0.4 percent and typical performance
   specifications would be nonlinearity error of ±0.5 LSB (least
   significant bit), temperature gain error of ±2 LSB, and temperature offset error of ±1 LSB. This is a total error of ±3.5 LSB,
   or 3.5/256 (±1.4 percent) for the A/D converter.
*Total Signal-Conditioning Error*:  Adding the amplifier error to
   the A/D converter error gives ±1.8 percent error, leaving
   ±1.2 percent error for the sensor. A general-purpose sensor
   with a maximum error of ±2 percent over temperature can be
   purchased at a reasonable price, but a sensor with a total error
   of ±1 percent or less would cost three to four times more.

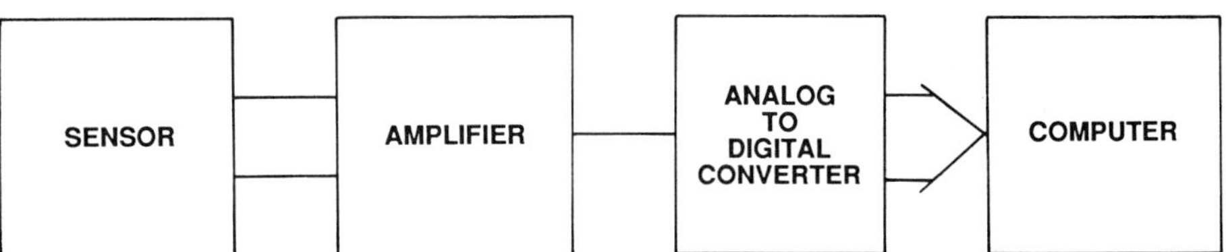



FIGURE 8.1   Pressure-measuring system.

*Options*:   The A/D converter could be changed to a 10-bit A/D
   converter, which increases the resolution to 1/1024, with linearity
   error of 0.5 percent, gain temperature error of ±4 LSB, and
   offset temperature error of ±1 LSB. The A/D error is reduced
   to 5.5/1024, or 0.5 percent. This will result in a total error
   for the A/D converter and the amplifier of 0.9 percent and
   leave ±2.1 percent for the sensor. Upgrading the A/D converter
   from 8-bit to 10-bit will much less expensive than trying to
   find a sensor with less than 1 percent total error.

### 8.3.4   Accuracy Calculations

Accuracy performance is directly tied to unit price: A sensor with
0.1 percent error is going to cost much more than a device with
2 percent error. The measurement accuracy required must be determined before a sensor can be selected. Today, 1 percent accuracy
is available for tens of dollars, depending on the quantity used;
in very large quantities, sensors can be purchased for less than
ten dollars. Most applications can use 1 to 2 percent static accuracy,
which is the sum of linearity, repeatability, and hysteresis.
    Some manufacturers specify a static accuracy that is the accuracy
at a reference condition (usually 25°C), but it is left to the user
to read the specifications and calculate the accuracy over the entire
temperature range. The user is interested in percent of reading
accuracy; for example, if 10.00 psi is applied to a pressure transducer, how accurately will the transducer measure the applied pressure? A realistic approach for calculating the total error due to
a number of independent errors is the root-sum-square (RSS) method.
This technique will provide a typical error rather than a worst-case
error. Simply adding all the worst-case errors puts an unrealistic
burden on the sensor and may result in the purchase of a sensor
that is more expensive than necessary. To find the RSS error,
each error is squared and added together; the square root of the
sum of all the errors is the typical error.

*Example*: Assume a sensor with the following specifications:

Offset: ±2 mV
Sensitivity: 100 mV ± 5 mV
Linearity: ±0.5% FS
Hysteresis: ±0.2% FS
Repeatability: ±0.1% FS
Offset TC: ±1% FS
Span TC: ±1% FS

If the offset and sensitivity are trimmed in the system, they are not summed with the other errors. If there is no adjustment offset and span errors are converted to percent FS (FS = 100 mV) and squared:

$$\text{RSS error} = (\text{Lin})^2 + (\text{Hyst})^2 + (\text{Rep})^2 + (\text{Off TC})^2 + (\text{Sp TC})^2$$

$$= (0.5)^2 + (0.2)^2 + (0.1)^2 + (1)^2 + (1)^2$$

$$= \pm 1.52 \text{ percent}$$

Simply adding all the maximum errors gives a maximum error of ±2.8 percent FS—almost twice the RSS error.

## 8.3.5 Electrical Specifications

Electrical specifications include parameters such as output signal, excitation, temperature range, and calibration. These must be compatible with the end-product specifications.

*Output Signal*

A number of sensor output configurations are available:

1. Analog voltage
2. Frequency
3. Digital
4. Current

Analog voltage and current output (4 to 20 mA) are the most common configurations for high-level-output transducers. Frequency and digital outputs are available on some higher priced devices.

Sensors are generally available with low-level outputs; differential analog voltage is the most common.

## Excitation

The two most common types of sensor excitation are constant current and constant voltage. A voltage reference would be added to use constant voltage, while constant current sources will have to be designed to use constant-current devices. A review of the difference between constant current and constant voltage may be useful at this point.

### Constant Current

- Maintains a constant current over a wide voltage range.
- Sensor bridge impedance changes as a function of temperature, providing span compensation.
- A precise current source must be provided to make highly accurate measurements.

Constant-current devices have the advantage that the sensor chip is used to sense the temperature and provide temperature compensation. If the medium to be measured is at a temperature different from ambient temperature, this can be a significant source of error. The other consideration is that the current source itself can be a potential error source. If the current changes, the bridge voltage will change proportionally, resulting in a system error.

### Constant Voltage

- Maintains a constant voltage over a wide current range.
- The bridge alone does not determine the span temperature compensation.
- The voltage reference device will determine the span accuracy.

A constant-voltage device requires a precise and stable voltage source to power the sensor. A Wheatstone bridge sensor is ratiometric, and any change in bridge voltage will change the sensor output proportionally.

Constant-voltage and constant-current sensors are not interchangeable. The temperature-compensation thermistors on a constant-voltage device will have no effect on the bridge voltage if the sensor is excited with a constant current source, and a constant-current sensor will not function properly powered from a constant voltage source.

In some applications, there may not be enough voltage available to use the standard temperature-compensation schemes. If the system is to be powered from low-voltage supplies such as 3 or 5 V, it may make more sense to put the full voltage across the bridge and change the gain in the amplifier by +2200 ppm/°C by using thermistors or other temperature-sensitive resistors to do the span temperature compensation.

*Transducer Excitation*

Amplified and calibrated transducers are available in two configurations: ratiometric or regulated.

*Ratiometric*:   Ratiometric transducers have the sensor and amplifiers tied to the power supply. As the power supply is changed, the transducer sensitivity and offset change proportionally (Figure 8.2). This can be useful if the sensor is used with a ratiometric analog-to-digital converter. The A/D reference and the sensor supply voltage are tied together and will automatically compensate sensitivity errors due to any changes in the reference voltage (Figure 8.3). This allows the device to be operated from a battery and automatically compensates for the gradual decrease of the battery voltage.

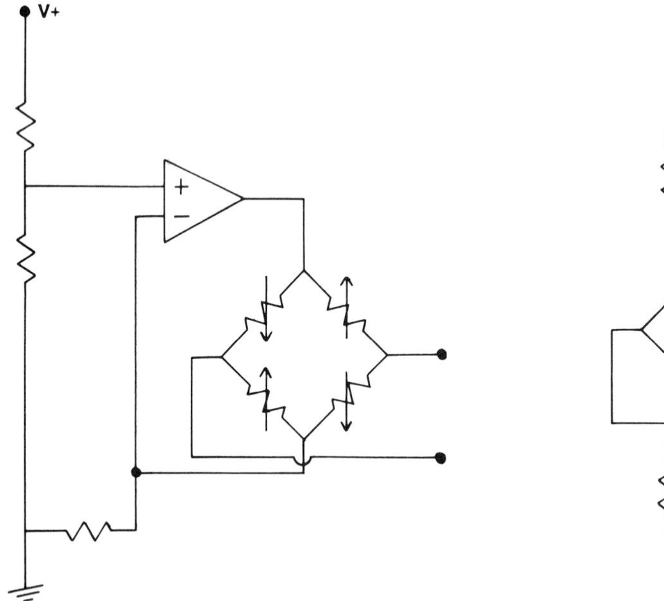


FIGURE 8.2   Ratiometric sensor.

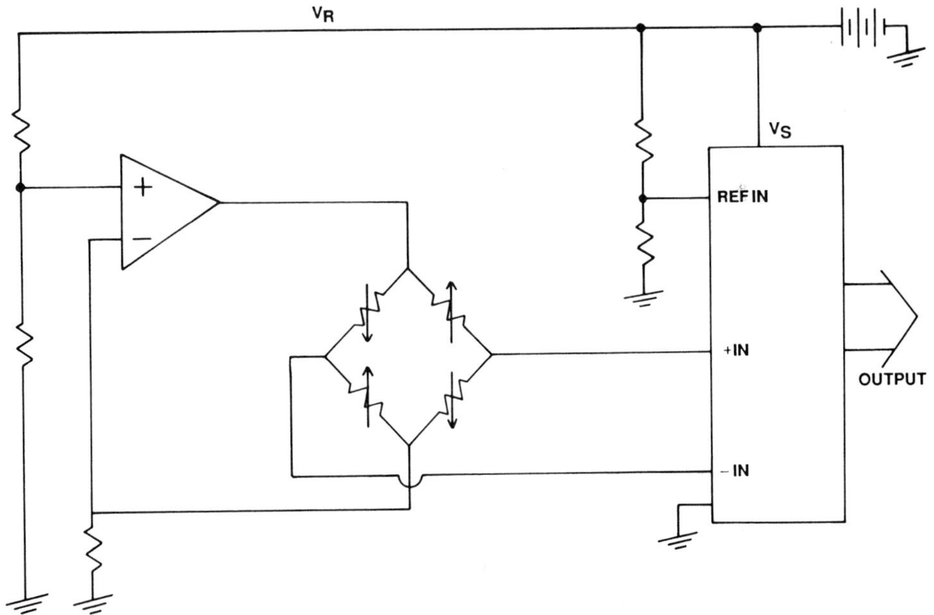

FIGURE 8.3   Ratiometric sensor and A/D converter.

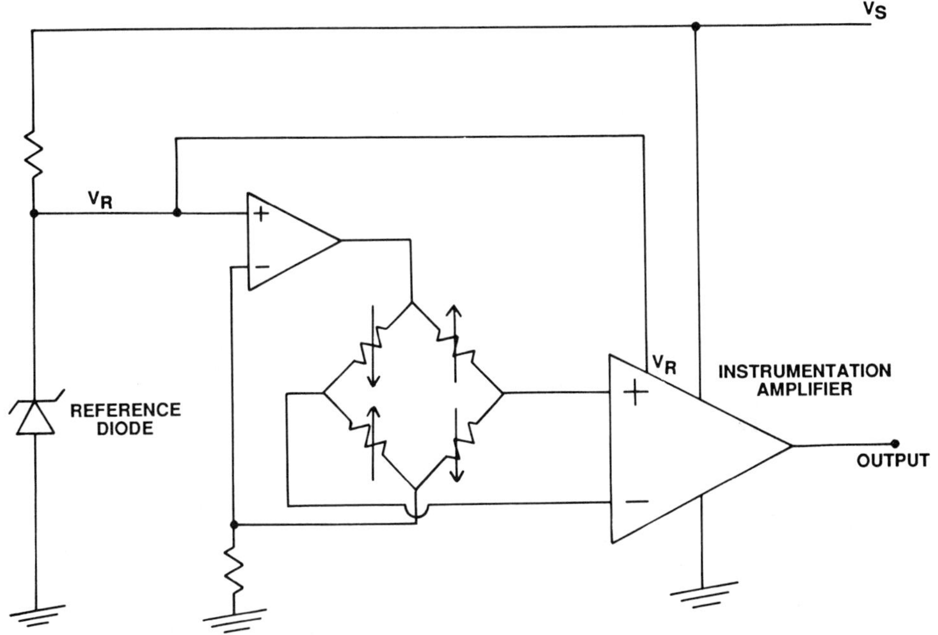

FIGURE 8.4   Regulated transducer.

*Regulated*:   Regulated transducers have internal reference voltages
and are independent of the power-supply voltage (Figure 8.4).
As long as the supply voltage is within the specified voltage
limits, the output will remain constant.

*Power Requirements*

Most applications use standard power supplies and power is not
a problem. In battery-operated devices like radio sondes or buoys,
where it is not easy to change the battery, long battery life and
special sensors are required. These devices need a low-power sensor
that has good repeatability specifications and fast warm-up. The
system is powered down most of the time; when a reading needs
to be taken, the power is applied, the measurement is taken, and
the system is returned to the power-down state.

## 8.4  TEMPERATURE REQUIREMENTS
### 8.4.1  Temperature Specifications

In what temperature range will the system be used? For systems
that are used in doors in an air-conditioned environment, there
is little concern with temperature; for systems used to monitor
boilers or to monitor pipeline pressure in Alaska, it is important
to ensure that the sensors used can operate over the required
temperature range.

All sensors and transducers are specified and calibrated over
a specific temperature range.  The three standard temperature ranges
are

- Commercial: 0°C to 70°C
- Industrial: -25°C to 85°C
- Military: -55°C to 125°C

The majority of today's applications can use standard commercial
or industrial temperature-range transducers. Only a few applications
like cryogenics or deep space need extreme-temperature operating
specifications. Here again, the application will determine the operating
temperature range. If the device is going to be used indoors, 0°C
to 70°C (commercial) is sufficient. If the device is used outdoors,
-25°C to 85°C (industrial) will cover most areas. The third standard
temperature range is military, -55°C to 125°C.

If the application requires a temperature range outside the
standard ranges, a special transducer has to be purchased or the
transducer has to be protected from the extreme temperatures by
using a heater and insulation.

## 8.4.2   Electrical Temperature Considerations

The temperature characteristics of the die and the materials used
in the package and sensing die will determine the temperature range
of the device. Factors that will have a direct effect on the operating
temperature range of the device are (1) leakage currents, (2) resistor
temperature characteristics, and (3) temperature-compensation technique.

### Leakage

Junction-isolated sensors have back-biased diodes that isolate the
bridge resistors from the substrate. The back-biased diode has
a small leakage current for each resistor (Figure 8.5). This leakage
current is temperature-dependent and will double approximately
every 8°C. A device that has 5 nA of leakage current at 25°C will
have a leakage current of 20 $\mu$A at 120°C. To operate a sensor
at 150°C and above, these leakage currents must be considered,
and the chip will require some basic design changes.

### Resistor Temperature Characteristics

The span-temperature characteristics of the sensors are directly
affected by the temperature characteristics of the bridge resistors.
Figure 8.6 shows the typical span and resistor temperature curve
of a piezoresistive sensor at constant bridge voltage. As shown
on the graph, these are both nonlinear functions, which makes
temperature compensation over a very large temperature range difficult. The slope and shape of the resistor temperature characteristics
are important when considering the sensor temperature compensation.
At very low temperatures, the slope of the resistor temperature
characteristics will decrease to 0 and change sign (see Figure 8.7).

### Temperature-Compensation Technique

The standard constant-current and thermistor temperature compensation work well over a limited temperature range, but these techniques
are not that easy to use over extended temperature ranges, due
to the difficulty in matching two nonlinear functions over a large
temperature range. If extreme temperatures are required, the sensor
must be compensated at the working temperature (if possible), or
an alternative temperature-compensation technique must be developed.

## 8.4.3   Mechanical Temperature Considerations

There will also be mechanical limitations, such as (1) package material, (2) sensor chip-mount material, (3) electrical connectors, and
(4) pressure seals.

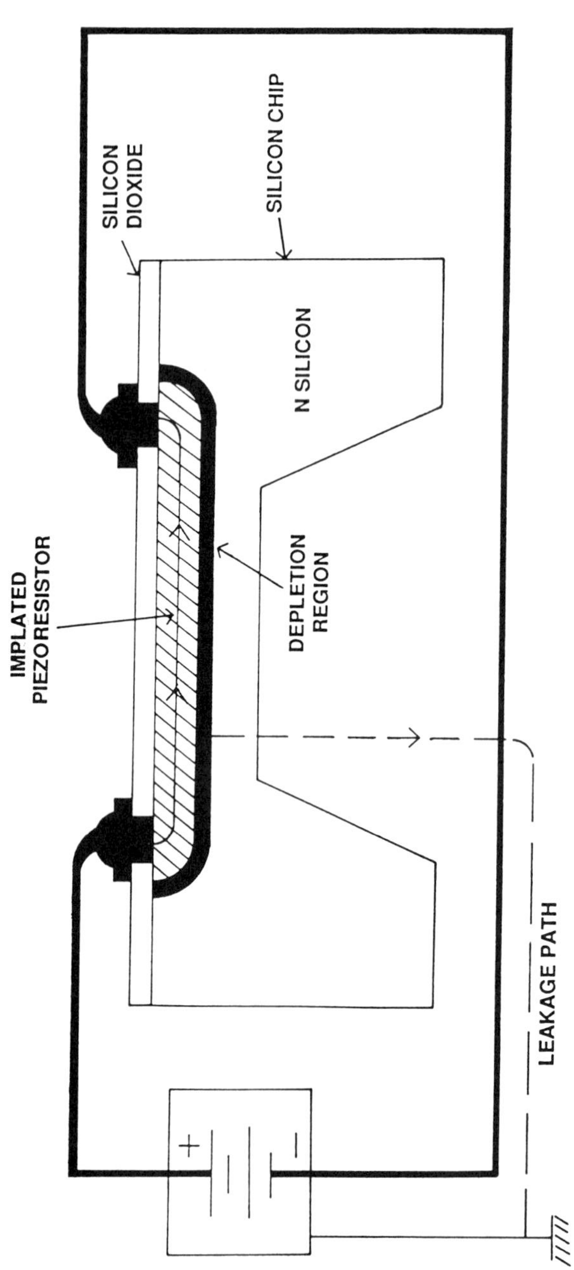



FIGURE 8.5  Junction-isolated sensor.

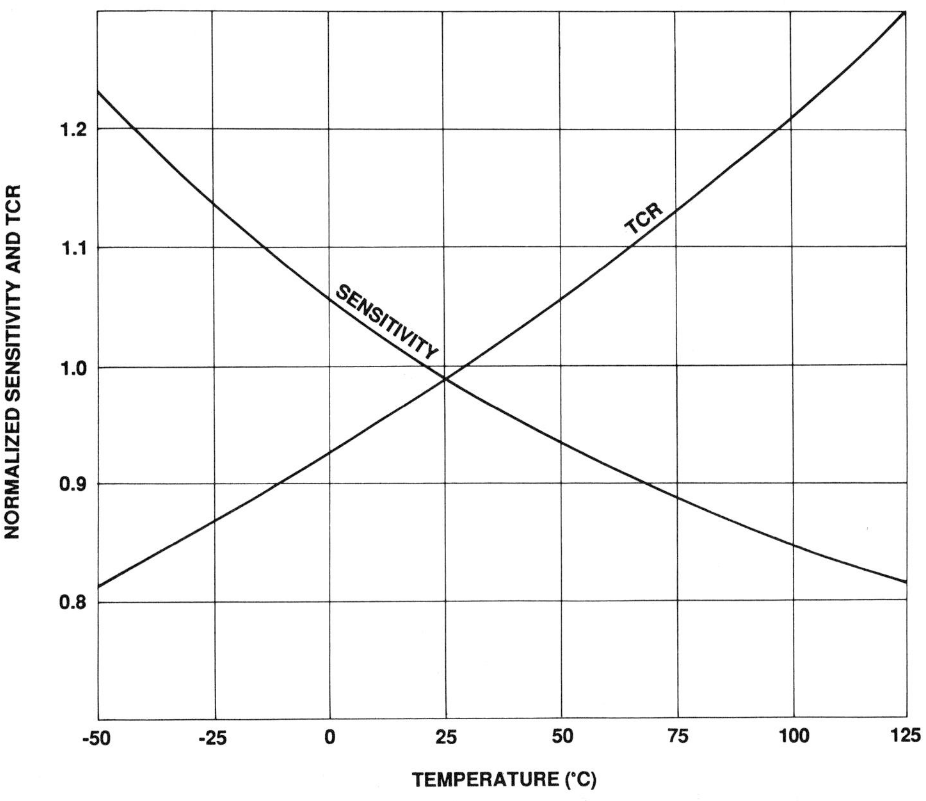


FIGURE 8.6   TC sensitivity vs. TCR.


*Package Material*

The most common package materials are stainless steel, ceramic,
and plastic. Stainless-steel and ceramic packages are used for high-
performance devices, while plastic is used for high-volume, low-cost
devices. Most ceramic and stainless-steel packages are good from
-40°C to 125°C, and the low-cost plastic package is used from 0°C
to 70°C or 80°C. Another consideration is pressure range. If an
application is to measure 10,000 psi, the package has to be stainless
steel or some stronger material.

The other packaging consideration when selecting a pressure
transducer or sensor is the material used to seal the package and
to mount the die to the header. If the sensor is potted with epoxy,
or the electrical connector is attached to the body with epoxy,
this can be a consideration in the overall performance of the device

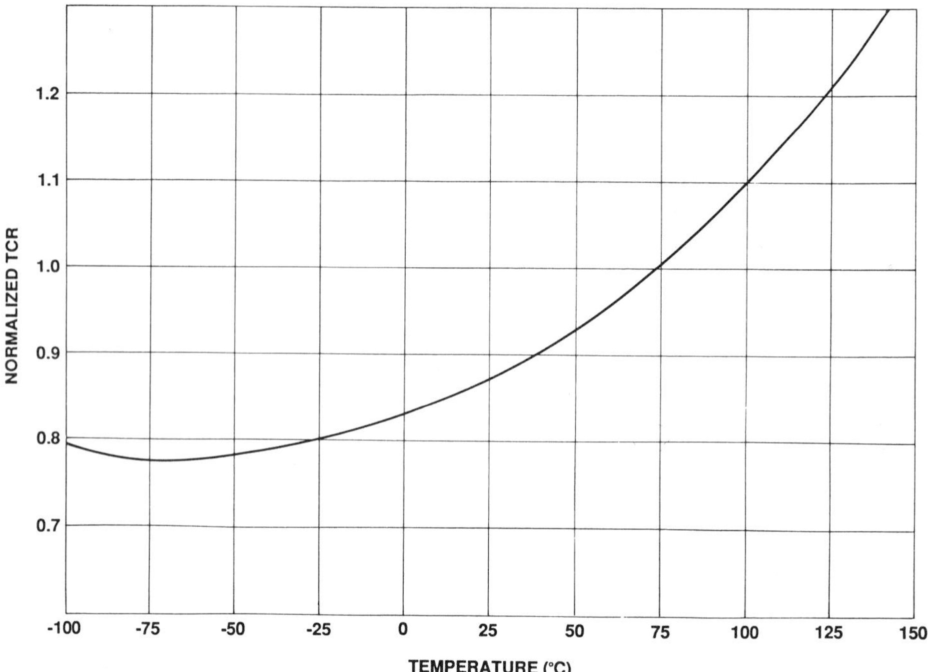

FIGURE 8.7  Bridge TCR.

over extended temperature ranges. A welded stainless-steel package
can be used over extended temperature ranges and is more reliable.

*Sensor Chip-Mount Material*

The material used to mount the sensor chip has to perform two
functions. In gage and differential sensors, it has to support the
chip, as well as make a pressure seal to the housing. RTV, epoxy,
and gold eutectic are the most common mount materials. RTV has
two advantages: It will not transmit strains to the chip from the
package strains or as a result of thermal expansion mismatch, and
it will remain compliant over a wide temperature range. The only
drawback to using RTV is that it does not bond to all materials
(some plastics), and it is not compatible with some media, such
as gasoline and silicon oil.

Epoxy is used in some sensors; it is stronger than RTV and
is compatible with gasoline and silicon oils. The drawback to using
epoxy is that it cures rigidly and can induce strain errors into
the sensor. Epoxy is generally used in higher pressure sensors.

Gold eutectic is another means of mounting the chip. A gold preform is melted and forms a eutectic with silicon. Gold is inert, is compatible with most media, and provides a hermetic seal at the silicon-package interface. The main drawback to the gold-eutectic mount is that the sensor chip constraint wafer must be silicon and the package material must have a metallic mounting pad that will wet with the gold. Plastic packaging cannot be used, due to the high temperatures (eutectic $\simeq$ 425°C) and inability to plate. Eutectic is used with ceramic and metal packages.

The mount used with the sensor can have a direct effect on the maximum temperature range and the media compatibility of the sensor. Eutectic will work over the largest temperature range, followed by RTV and then epoxy.

### Electrical Connections

Two sets of electrical connections must be made in the sensor. First, the sensor chip must be connected to the package connector; second, the package connector must make connections to the external circuit. Piezoresistive sensors are basically semiconductor devices and have the same electrical-connection techniques; ball bonds or compression bonds to pure aluminum pads are the standard for semiconductor devices. Thin gold wire is used to bond from the sensor to the package connector. Gold has a low melting temperature and is compatible with aluminum. This technique makes a good electrical contact and gives a low-impedance connection over wide temperature ranges.

Unfortunately, the other electrical connection will not work over extended temperature ranges. Mechanical connectors become intermittent at extended temperatures, and solder will reflow at temperatures much over 200°C. A high-temperature solder can be used to extend the temperature range to 250°C to 300°C.

### Pressure Seals

The pressure seals used in the package also determine the temperature range over which the sensor can be used. The most common pressure seals are O-rings, epoxy, barb fittings, pipe fittings, and welds.

*O-Rings*  O-rings are the most common pressure-seal technique. They are doughnut-shaped rings made of some compliant material. There are many kinds of O-ring seal configurations: static seals (stationary), dynamic seals (pistons), and face seals, for example. As shown in Figure 8.8, a groove (or gland) slightly wider than the O-ring is cut into the piece to be sealed; the seal is made when the pressure pushes the O-ring against the side of the groove and

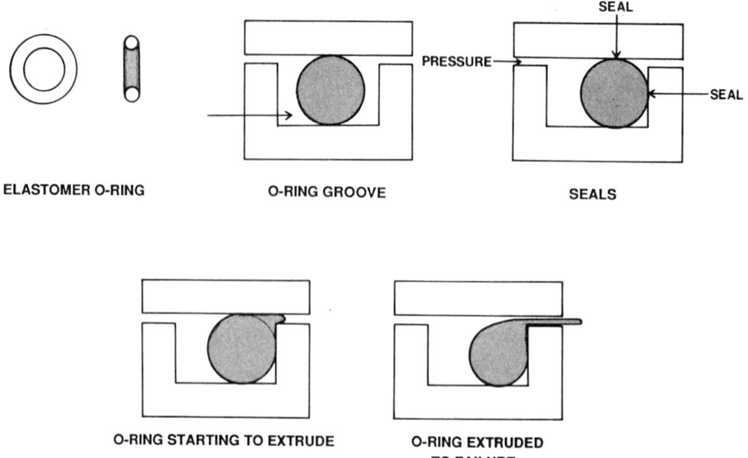


FIGURE 8.8   O-ring seal.

the joining piece. The O-ring is free to move back and forth in
the groove as pressure is applied. The dimensions of the groove
are critical: If it is too small, the O-ring fills the groove and cannot
move; if it is too large, the O-ring does not reach the joining piece
and no seal is made.

The O-ring material determines the operating temperature range
of a particular O-ring. O-rings are available in dozens of materials,
each one designed for particular applications. Some materials are
compatible with oils and hydraulic fluids, but not water and gases;
others are formulated for high temperatures or special fluids like
freons.

In applications that use O-ring grooves, the compliant O-ring
is free to move in the groove and rolls as the pressure is applied.
If it becomes still or brittle due to temperature or aging, it will
not seal properly. As shown in Figure 8.8, the O-ring seals on
two surfaces; as pressure is applied, the O-ring is forced to one
side of the groove. As more pressure is applied, the soft O-ring
is forced into the gap between the two surfaces. The maximum
gap and durometer hardness of the O-ring must be calculated for
the maximum working pressure of the system. If too much pressure
is applied, the O-ring is extruded through the gap and the seal
is lost. Most low-temperature applications use silicon materials,
because of their ability to remain compliant at low temperatures.

As shown in Figure 8.9, an O-ring can be used to seal two
round surfaces or two flat surfaces. In some face-seal applications,

the O-ring is squeezed into a small groove and used more as a
compliant washer to make the pressure seal. O-rings are used over
the full range of pressures, from 10 to 10,000 psi. The higher
the working pressure, the more critical the O-ring groove dimensions
and material hardness.

*Epoxy*   In some low-pressure sensors, epoxy or RTV may be used
to seal the sensor chip to the package or to pot the electrical connections. Some manufacturers also use epoxy and RTV to seal some of
the plastic packages, using the epoxy as an adhesive to hold the
package together.

*Barb Fittings*   Most low-pressure sensors and transducers employ
barbed pressure ports or pressure tubes that are used with compliant
hoses. The hose is simply pushed over the tube with no clamps
(Figure 8.9). This technique will work for pressures of 1 to 20 psi.
At higher pressures, the tube will become soft and will come loose
at higher temperatures.

   The compliant tube can be used at higher pressures if a hose
clamp is added. With a hose clamp, the limit will be the strength
of the hose material.

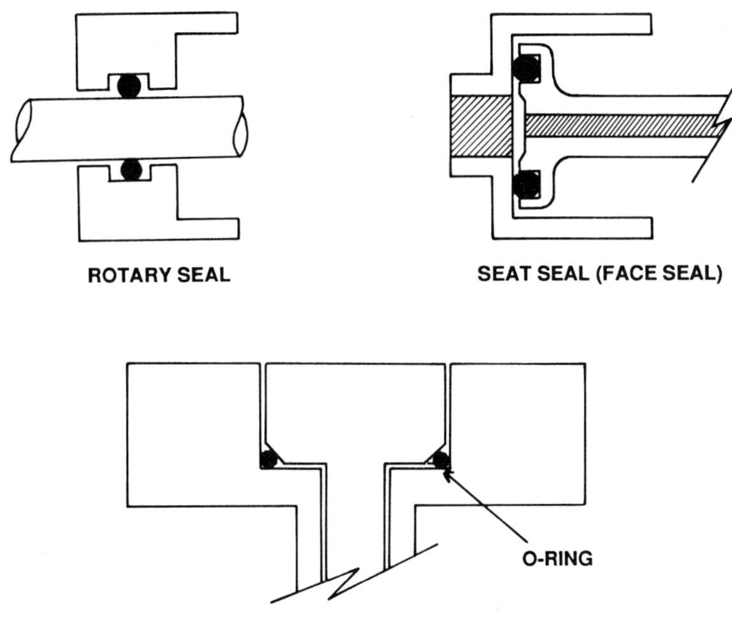


FIGURE 8.9   O-ring applications.

*Pipe Fittings*  Pipe fittings are used on most higher pressure sensors ($\geq$100 psi). There are a number of pipe threads available; NPT (national pipe thread) is probably the most common. Some care should be taken to ensure that the sensor thread is compatible with that used in the equipment. Pipe threads give a good seal over extended temperature ranges.

*Welds*  Welds provide the most reliable pressure seals, because they are not affected by temperature extremes and are not sensitive to the measuring medium. The weld gives a hermetic seal and is the preferred pressure seal where leaks would be hazardous or costly. Unfortunately, to be welded, the case material must be steel, which increases the price of welded sensors.

All the parameters discussed above will have a direct effect on the operating temperature range of a sensor or transducer.

## 8.5  PRESSURE RANGE

The system designer has two options when selecting a pressure sensor or transducer for a specific application: A sensor can be custom made, which will be more costly, or a standard product can be bought off the shelf, which will compromise some specifications to get lower cost.

One of the most common specifications that must be compromised is pressure range. Most transducer manufacturers have standard products available in specific pressure ranges. The most common ranges are

0-1 psi
0-5 psi
0-10 psi
0-15 psi
0-30 psi
0-50 psi
0-100 psi
0-300 psi
0-500 psi
0-1000 psi
0-5000 psi

If the designer's working pressure range happens to be between the standard ranges, there are two options: overpressure the lower pressure range, or use some percentage of the higher pressure range. The designer must review the specifications of each sensor and try to determine what effect operating the device at pressures

above and below the specified pressure range will have on the overall accuracy. One problem is that the sensor is specified as a percentage of the full scale or specified pressure range. Thus, a sensor
with a 1 percent FS offset-TC error at the specified pressure of
10 psi will have a 2 percent error if used with a working pressure
of 5 psi. As a rule, errors that are constants (do not change as
a function of pressure)—such as offset parameters—become larger
with reduced working pressures, while parameters that are dependent
on pressure—such as linearity and span parameters—decrease when
used at lower working pressures.

Key piezoresistive sensors will vary as follows:

| Parameter | Relationship to working pressure |
| --- | --- |
| Offset TC | Varies inversely to working pressure |
| Span TC | Varies proportionally to working pressure |
| Linearity | Varies proportionally to working pressure |

Another parameter that must be considered is maximum overpressure of the sensor. As a percent of working pressure, the
maximum overpressure of a sensor that is used below its specified
pressure range increases significantly. In some applications that
require a large overpressure specification, a designer may be forced
to use a sensor at a reduced pressure range to get the overpressure
range required.

The most common solution is to try to measure small pressures
(5 to 10 in. water) with a 1- or 5-psi sensor. Very-low-pressure
sensors are relatively expensive, while 1- and 5-psi sensors are
more reasonably priced. If the offset errors can be tolerated or
corrected, small pressures can be measured.


## 8.6 MECHANICAL

The size and weight of the sensor can be a major concern in some
applications, such as aircraft and spacecraft, but in most applications,
the concern is if the device will fit in the overall package and if
the pressure connections are compatible with the rest of the system.
If it is a redesign of an existing product, the sensor must be exactly
the same size and have the same pressure connections as the device
currently used. The pressure ports and mounting holes must be
in the same location, so that the device can be installed in the
existing package. Finding a replacement sensor or transducer standard product is not likely; in most cases, a special device will have
to be designed.

New designs are less critical; if the package is a certain size or smaller, it can be used.


## 8.7  COST

The selling price of the sensor or transducer is not the only consideration when selecting a pressure sensor. Will capital equipment, such as ovens or precision pressure sources, have to be purchased? How much extra labor will the pressure sensor add to the overall calibration of the system? It comes down to the basic question: Is it less expensive to buy a sensor and add the temperature compensation and signal conditioning, or to buy a conditioned transducer?


### 8.7.1  Purchase Price

The purchase price will be directly affected by the following factors:

1. Specifications
   a. Accuracy
   b. Temperature range
2. Package
   a. Plastic
   b. Ceramic
   c. Stainless steel
3. Quantity

What is available to the user?

1. Sensor chip
   a. Electrically probed
   b. Not pressure-tested
   c. High-volume only
   d. Lowest unit cost
2. Packaged sensor
   a. Pressure-tested
   b. Not compensated
   c. Low unit cost
3. Temperature-compensated sensor (packaged)
   a. Tested
   b. Temperature-compensated
      (1) Compensation installed
      (2) Compensation values provided
   c. Moderate cost

4. Conditioned transducer
   a. Packaged
   b. Temperature-compensated
   c. Conditioned (high-level output)
   d. Most expensive

The unpackaged sensor chip is the least expensive, but the user must have semiconductor experience and chip-handling, mounting, bonding, and testing facilities.

The packaged sensor is ready to pressure-test and temperature-compensate. The temperature-compensated sensor and finished (high-level-output) transducer are the easiest to use and require minimum value added after purchase, but have the highest unit price.

## 8.7.2  Calibration and Compensation

*Conditioned Transducer*:  The conditioned transducer needs no additional calibration or compensation; power and pressure are applied, and the transducer is usable as purchased. The only consideration is that a standard pressure range is available that will be compatible with the application.

*Temperature-Compensated Sensor*:  The temperature-compensated sensor is more versatile, but requires signal conditioning and calibration. The output of the sensor can be amplified, converted to a current, or converted to a digital output. It can be calibrated for any pressure range.

*Packaged Sensor*:  The packaged sensor is less expensive, but requires more labor by the user. It is left to the user to temperature-compensate and calibrate each device.

*Sensor Chip*:  The least expensive sensor is the sensor chip. The wafer is processed and probed, marking all open and shorted bridges. The wafer is then sawed into individual dies and sent to the customer. It is left to the customer to package, temperature-compensate, and calibrate each sensor.

Each customer has to look at the application and make the build-or-buy decision. Can the customer temperature-compensate the sensor for less than the difference between the price of a basic sensor and a temperature-compensated sensor?

## 8.7.3  Capital Equipment

The unit price of the sensor is not the only consideration when making the build-or-buy decision. A number of expensive pieces of test equipment should also be considered:

1.  Temperature chambers
2.  Precision pressure source
3.  Pressure manifolds
4.  Test fixtures (electrical)
5.  Printed circuit boards
6.  Bonders

If special test equipment like pressure sources and temperature chambers are not in place, the cost of the additional equipment must be added to the total price of the sensor.


8.7.4  Maintenance

Some expensive transmitters can be repaired and recalibrated, but the trend is toward disposable sensors. In some medical applications, the sensor is used once and then thrown away with all the used plastic tubing; sterilizing the sensor is more expensive than replacing it. This is one of the high-volume applications that takes advantage of the high-volume processing of piezoresistive sensors.

# 9

# Auto-Zero and Auto-Reference Techniques



## 9.1 INTRODUCTION

In some applications, the performance of a medium-priced sensor
can be greatly improved by implementing a simple auto-zero or
auto-reference technique. As the name implies, an electronic circuit
can be designed to automatically adjust out any offset errors in
the sensor circuit. An auto-zero circuit forces the output to be
equal to a predetermined reference voltage on command. The auto
zero can be designed to zero the circuit at turn on, every time
the system is vented, or just before each measurement is made.

The microprocessor has introduced a degree of flexibility to
the signal conditioning of sensors that had not been available before.
The microprocessor can be programmed to control the sequence of
events, modify the sensor output, and store data to be used at
a later date. The microprocessor can be programmed to control
the auto-zero electronics, as well as the solenoids and valves, to
ensure that the reference pressure is present when the auto-zero
circuit is activated. The microprocessor can also be used to auto-
reference the measurement. By taking a reading at a reference
pressure and storing that number in memory, the microprocessor
can mathematically compensate for any offset errors in the system.

## 9.2 DEFINITIONS

*Auto Zero*

An auto-zero circuit is defined as a circuit that generates a signal
equal in magnitude and opposite in sign to the system offset errors

and holds that signal through the measuring sequence. The sequence of events is as follows:

1.  Reference pressure is applied to the sensor.
2.  The sign and magnitude of the output error are measured.
3.  A correction voltage is generated and summed in the electronics to force the output to zero
4.  The correction voltage is maintained by a sample-and-hold circuit.

*Auto Reference*

The auto-reference technique is defined as a system that measures the output of the sensor at a reference condition (usually zero) and stores that number in memory. As each subsequent measurement is taken, the stored offset number is subtracted to get the true reading.

As defined above, the auto-zero technique is performed by hardware and the output voltage is modified, while the auto-reference technique is performed in software and the sensor output is not altered. If the error is not large enough to saturate the output amplifier, either technique can be used.

The benefit of using auto zero or auto reference is that all errors associated with offset, such as offset temperature error, are reduced to zero, which improves the accuracy of the sensor.

*Offset Errors*

An auto zero or auto reference is generally used to correct for offset errors in the sensor or system. A typical specification for a 15-psi gage sensor is as follows:

Pressure range:   15 psig
Sensitivity:   6.67 mV/psi typical
Full scale:   100 mV typical
Zero-pressure offset:   ±2 mV
Linearity and hysteresis:   ±1 FS
Temperature effect on span:   ±1% FS
Temperature effect on offset:   ±2% FS
Repeatability:   ±0.5% FS
Input impedance:   5 k$\Omega$ typical
Long-term stability (span and offset):   ±0.2% typical

Looking at the worst-case error for the sensor, including an initial offset of ±6.7 percent of full scale, how much of this total error is directly related to offset?

Zero-pressure offset:   ±2 mV or ±2%
Temperature effects on offset:   ±2% FS
Long-term stability:   ±0.1% FS
Total:   ±4.1% FS

Out of 6.7 percent worst-case error, 4.1 percent is offset-related.
By reducing the offset errors to zero, a 6.7 percent sensor can
be improved to ±2.6 percent worst-case error. In some applications,
this means that a low-cost sensor can be used and still maintain
the system accuracy.


## 9.3  AUTO-ZERO TECHNIQUES

Auto-zero circuits can be analog or digital, depending on the application. Figure 9.1 shows the block diagram of a basic auto-zero
circuit. The output of the amplifier is sensed by the auto-zero
circuit and compared to the reference input, and a correction signal
is generated and summed with the sensor output to force the amplifier output to the same voltage as the reference. If a sample-and-
hold amplifier is used, a logic input is needed to control the state
(sample or hold) of the amplifier.

   Digital voltmeters have used basic auto-zero techniques for
some time. The input terminals are disconnected from the input
amplifier and the amplifier and A/D converter are automatically
adjusted for zero output.

   Some applications cannot use auto-zero techniques. For example,
long-term measuring of a varying pressure that has no reference
pressure in the operating cycle should not use auto zero.

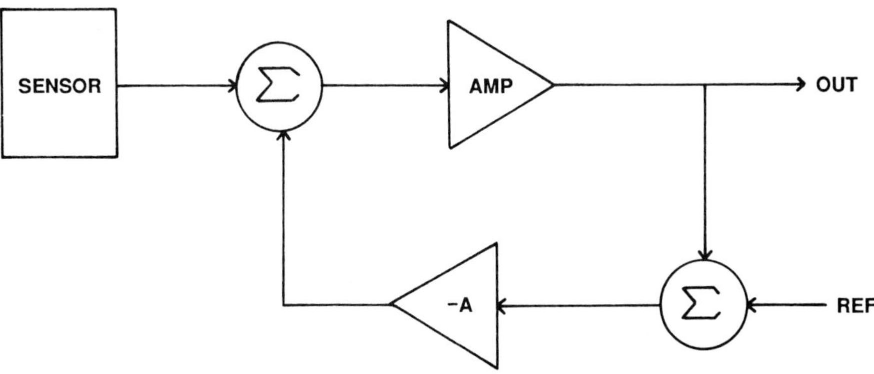


FIGURE 9.1  Basic auto-zero circuits.

Figure 9.2 shows the offset-temperature characteristics of a typical sensor. The 25°C curve is the static accuracy of the sensor, and the 0°C and 80°C curves show the effects of temperature on the sensor output voltage. The slope stays constant and the curve moves vertically. This error is a constant and will have the same effect at any point along the curve.

Figure 9.3 shows the effects of temperature on sensor span or sensitivity; again, the 25°C, 0°C, and 80°C characteristics are plotted. From the plot, it should be noted that the sensitivity error is a function of the applied pressure. At 0 psi, there is no error; as pressure is increased, the error increases until full-scale pressure is reached. This is important if a fraction of the full-scale pressure range is to be used.

The auto-zero circuit eliminates the offset error, leaving only the sensitivity error. This reduces the total sensor error by 50 percent minimum in most instances.

The auto-zero circuit can be used to zero out the offset errors in more than just the sensor and amplifier circuit. The error voltage

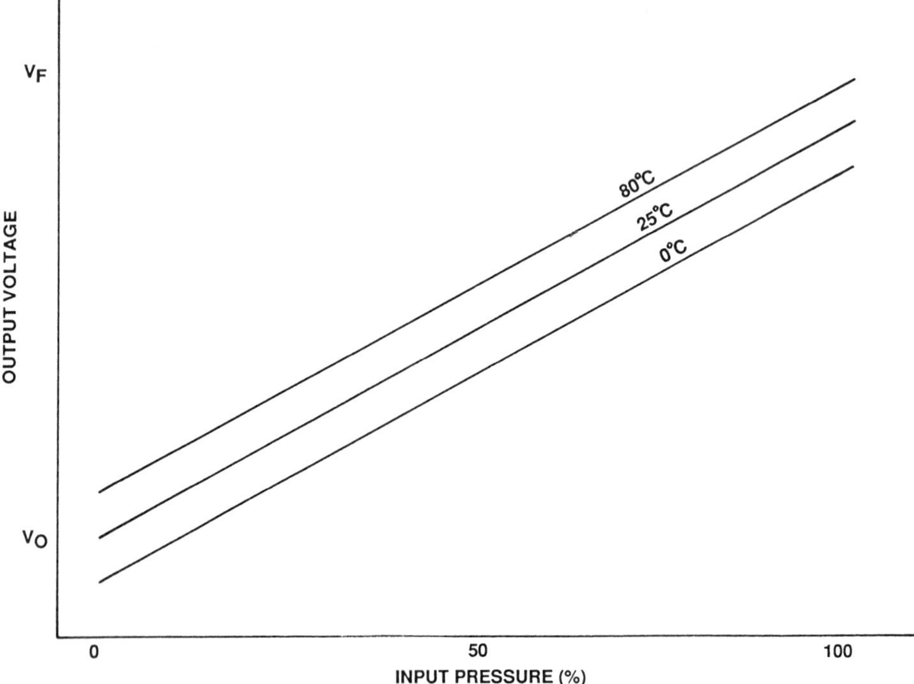


FIGURE 9.2   Offset-temperature characteristics.

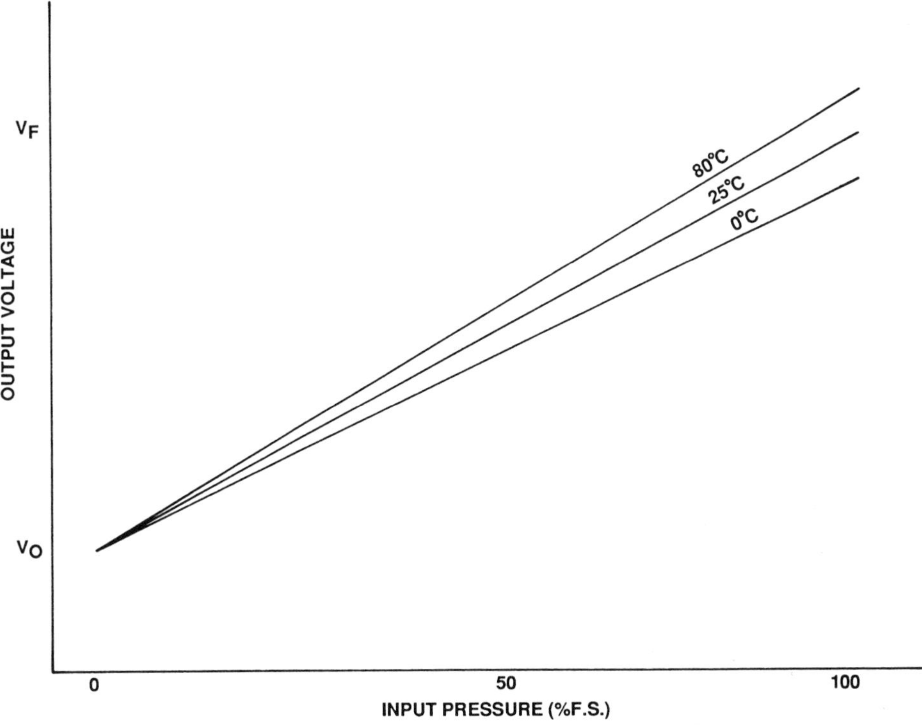


FIGURE 9.3   Temperature effects on span.

can be picked off after any stage in the circuit. To get the maximum benefit from the auto zero, as many stages as possible should be included in the auto-zero feedback loop. Using this technique will reduce total system errors.

### 9.2.1  Analog Auto-Zero Circuits

One of the simplest analog circuits can be made by putting an integrator in the common-mode feedback loop of an instrumentation amplifier. Figure 9.4 shows such a circuit. Amplifier $A_3$ has very gain at DC, and the gain rolls off at 6 dB per octave. The output of amplifier $A_2$ drives the reference input of the instrumentation amplifier, forcing the instrumentation amplifier output to equal the reference voltage at DC. From the earlier discussion on instrumentation amplifiers, the reference input can be used as an independent offset-adjust pin without affecting amplifier sensitivity. The reference

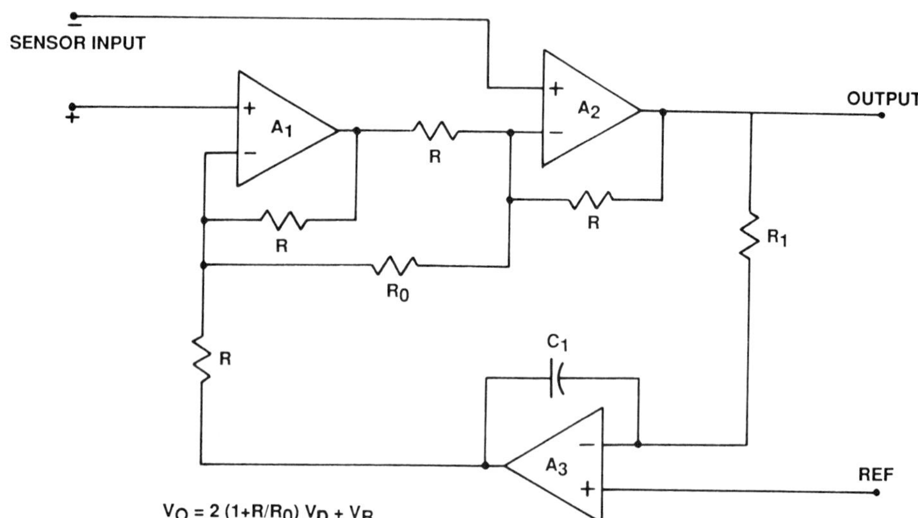



$$V_O = 2\,(1 + R/R_0)\,V_D + V_R$$

FIGURE 9.4  Active auto-zero circuit.

input pin has a gain of 1 to the output. The values of resistor $R_1$ and capacitor $C_1$ set the corner frequency of the circuit. When the reactance of the capacitor equals the resistor value, amplifier $A_3$ has a gain of 1 (Figure 9.5). At higher frequencies, the gain will be less than 1 and the output will respond to changes in the input voltage. Using the values specified in Figure 9.5, the feedback is unity at 5 Hz and 0.005 attenuation at 1 kHz. The amplifier will compensate for any DC offset in the system, but cannot respond fast enough to compensate for transients and higher frequency pressure changes. This circuit has limited applications, since it will respond only to pressure changes. The circuit needs no control logic and automatically compensates for any DC errors generated in the sensor or amplifier.

A second analog circuit uses a sample-and-hold amplifier to provide the correction voltage to the summing junction and allow the circuit to respond to DC. The sample-and-hold amplifier (Figure 9.6) has a high-input-impedance unity-gain amplifier, a switch to connect the amplifier to the input during the sample mode and disconnect the input during the hold mode, and control logic to turn the switch on and off. In the sample mode, the output of the sample and hold tracks the input; when the control input changes state, the output of the amplifier holds the last voltage on the hold capacitor. The circuit shown in Figure 9.7 is basically the same as the circuit shown in Figure 9.4, except that a sample-

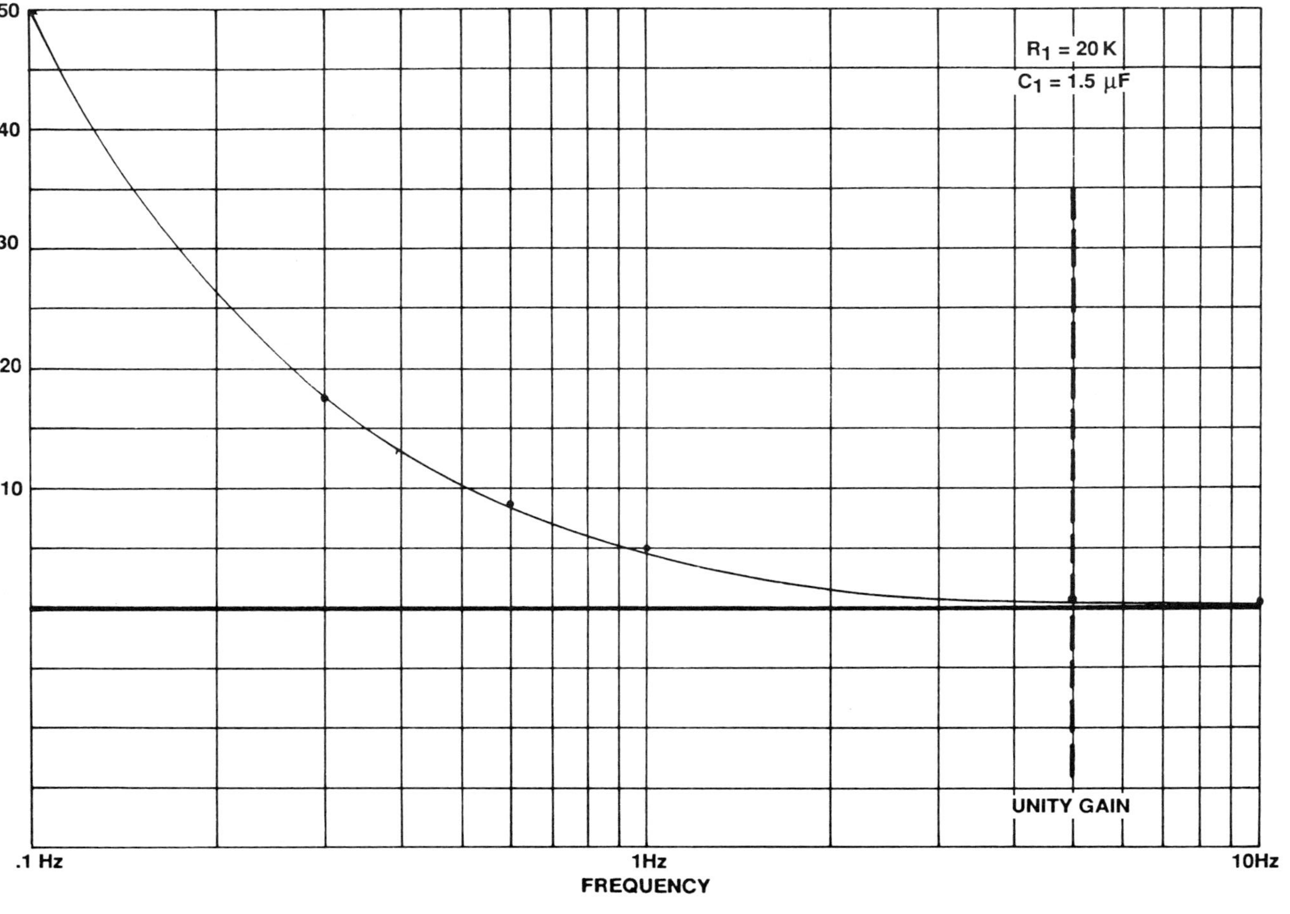


FIGURE 9.5  Amplifier $A_3$—gain vs. frequency.

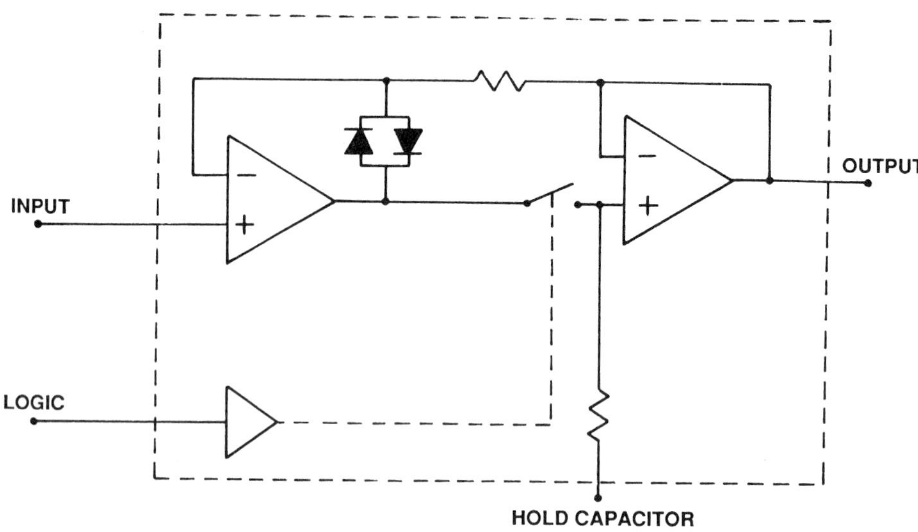


FIGURE 9.6   Sample-and-hold amplifier.

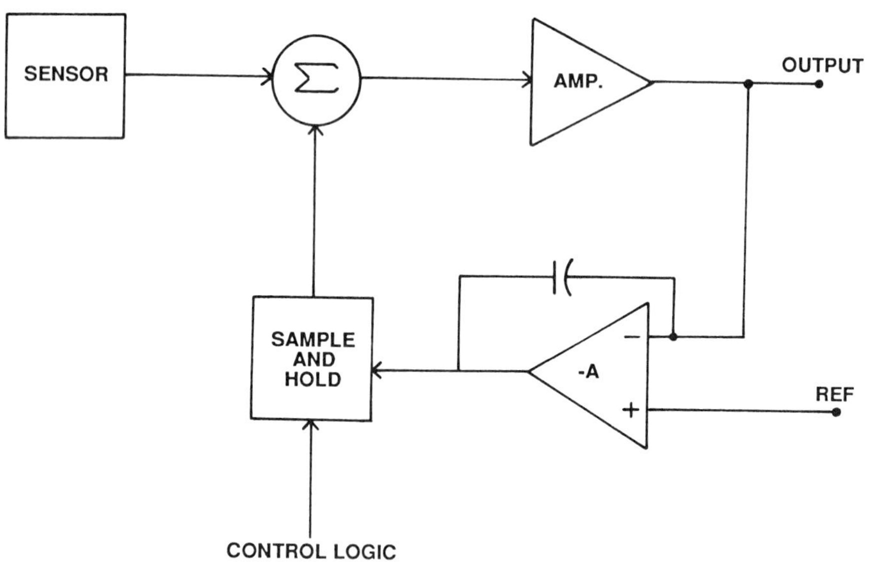


FIGURE 9.7   Sample-and-hold auto zero.

and-hold amplifier has been added to the feedback loop. In this circuit, the sample and hold is controlled by the mode input pin on the sample-and-hold amplifier. In the sample mode, the sensor amplifier output voltage is forced to be the same as the reference voltage by the sample-and-hold amplifier. In the hold mode, the last correction voltage is held by the sample-and-hold capacitor. The correction signal will slowly change or droop, depending on the leakage currents in the sample-and-hold amplifier. A good sample and hold will droop about 1 mV/25 ms. In a second, the voltage will change 40 mV. This circuit is useful for fast, repetitive measurements where the zero-correction voltage can be updated frequently. This circuit has limited uses, as well, due to the relatively short read cycles.

### 9.3.2　Digital Auto-Zero Circuits

Digital circuits can also be used to zero the output of a sensor. A D/A converter can be used as a digitally controlled sample and hold. The basic digital auto-zero circuit is shown in Figure 9.8. Clock pulses advance a binary counter, which provides the digital

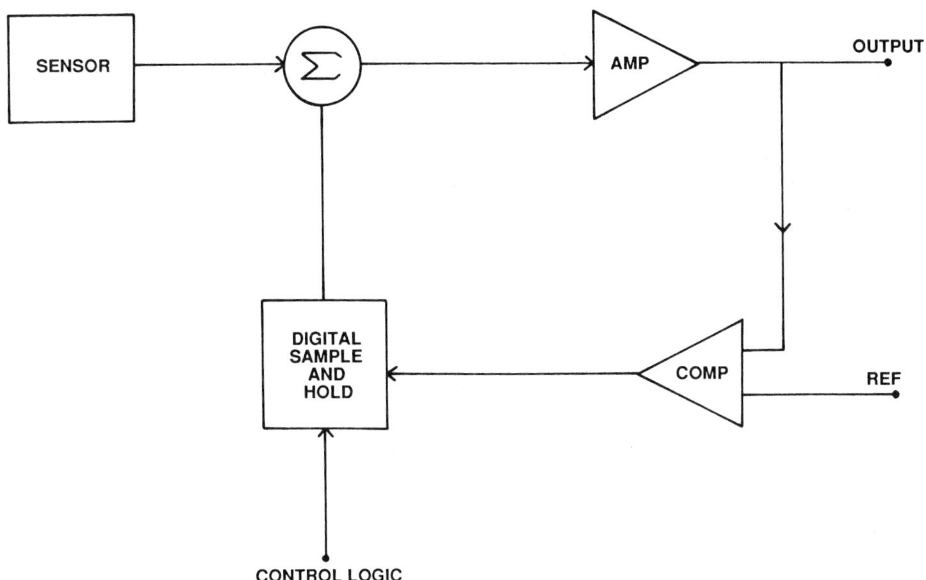


FIGURE 9.8　Basic digital auto zero.

input to the D/A converter. Each clock pulse increases the output voltage by one least significant bit. The output of the D/A is a staircase or ramp voltage; the D/A output is summed with the sensor output. The amplifier output is monitored by a comparator and, when the amplifier output equals the reference input, the comparator changes state, inhibiting the clock pulses and putting the circuit into the read mode. The circuit will stay in the read mode until another zero command is encountered.

### 9.3.3  Digital Sample and Hold

A D/A converter is a digitally controlled voltage source. If the digital input remains the same, the output voltage will remain fixed indefinitely (infinite sample and hold). The sample and hold consists of a D/A converter and a counter or buffer to maintain the digital input to the D/A, a clock, and a logic circuit to reset the counters and inhibit the clock. This is a big advantage over the analog approach, because the correction signal can be held for hours or even days. The number of bits in the D/A converter (DAC) determines the resolution of the correction circuit. An 8-bit DAC divides the output into 256 equal voltages, a 10-bit DAC divides the output into 1024 equal voltages, and a 12-bit DAC divides the output into 4096 equal voltages. The compromise is that the more bits in the DAC, the slower the zeroing sequence due to the additional counts. The more bits in the DAC, the closer the output will be to the reference voltage.

Figure 9.9 shows the D/A converter and binary counters used as a digital sample and hold. When the control logic is a zero gate, the logic circuit sets the counters to zero, creating a large offset at the output. Clock pulses are then fed to the counters, creating a staircase voltage at the analog output. The counters are allowed to count until the output voltage equals the reference voltage and the comparator changes state. The control logic senses the comparator output, inhibits the clock pulses to the counters, and puts the system into the read mode until the next zero command is encountered.

Figure 9.10 shows a CMOS binary counter and an R-2R ladder network that are used to perform the same function. The counter acts as the latch, as well as driving the ladder network. The output of the CMOS counter swings from ground to power-supply voltage. The supply voltage determines the output voltage range of the D/A converter by acting as the reference voltage. In this application, the value of the correction voltage is not important, but the stability of the reference voltage is critical. If the reference voltage moves, the DAC output will move proportionally. Since the supply or reference

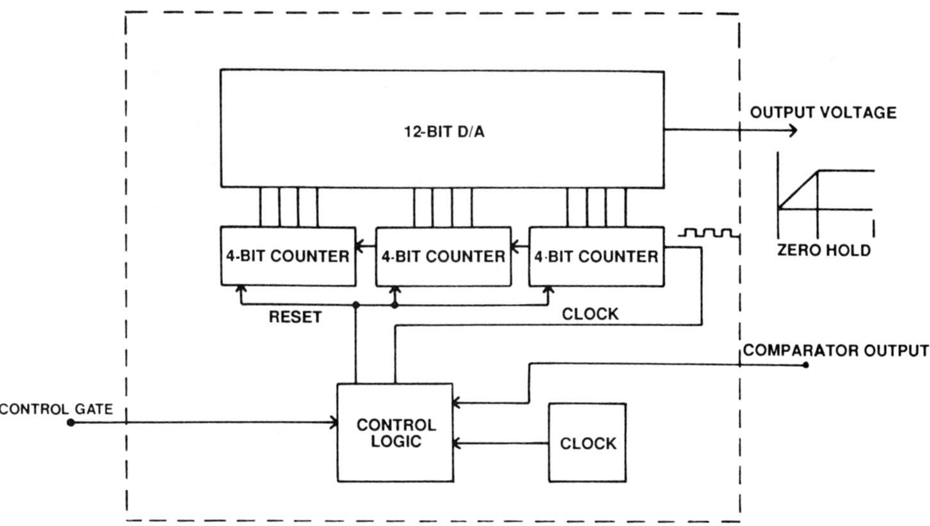


FIGURE 9.9  Digital sample and hold.

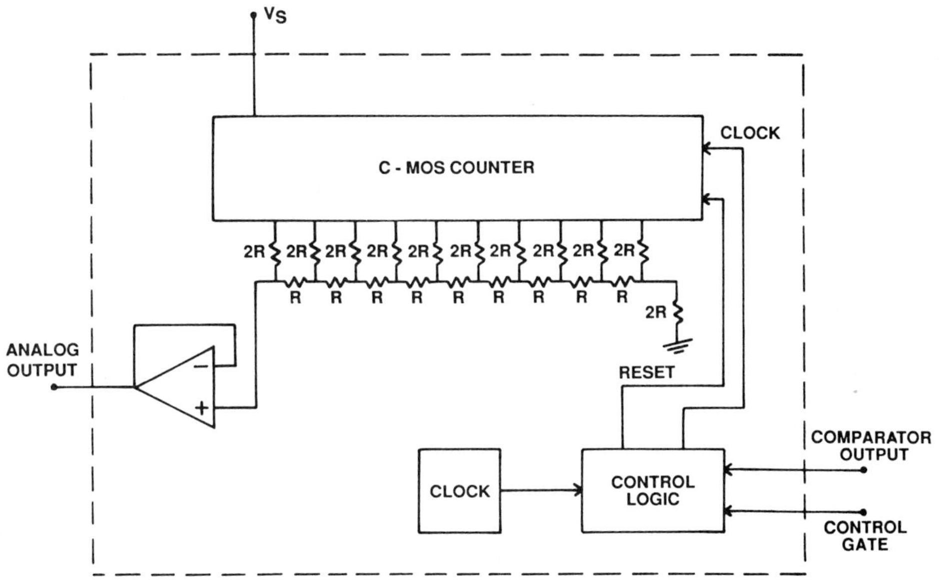


FIGURE 9.10  CMOS counter sample and hold.

voltage determines how stable the auto-zero circuit is with time, a voltage-reference device or good voltage regulator should be used. The CMOS counter will not drive heavy loads (low-impedance loads), which forces the R-2R ladder network to be relatively high impedance ($R \simeq 50$ k$\Omega$). This approach is less expensive, due to lower part count and generally less expensive parts.

Both circuits require some digital logic circuits to control the auto-zero process (Figure 9.11). A flip-flop can be used to buffer the comparator output to ensure that clock pulses won't be fed to the counter every time the comparator output is in the high state. The flip-flop output can also be used as a busy signal. A simple NAND gate can be used as a switch to control the clock pulses. The logic circuit also contains a one-shot to zero the counters and initiate the auto-zero sequence.

The auto-zero technique is very versatile. There can be many variations of the basic circuit: It can be used as a digital scale, a circuit can be added to automatically zero any voltage less than the reference, or a faster circuit can be designed to track the error.

### 9.3.4  Digital Auto-Zero Applications

*Automatic Blood-Pressure Tester*

A blood-pressure tester (sphygmomanometer) is an application that is perfectly suited to an auto-zero technique. Just before each test, the cuff is deflated, giving a reference condition (atmosphere) at which the sensor can be zeroed. Then, the cuff pressure is increased to approximately 200 mmHg and slowly deflated. As the pressure decreases, the systolic and diastolic pressures are measured

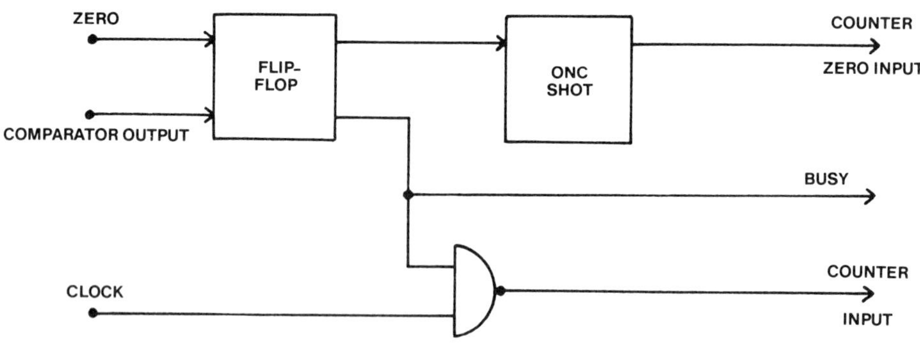


FIGURE 9.11   Control logic circuit.

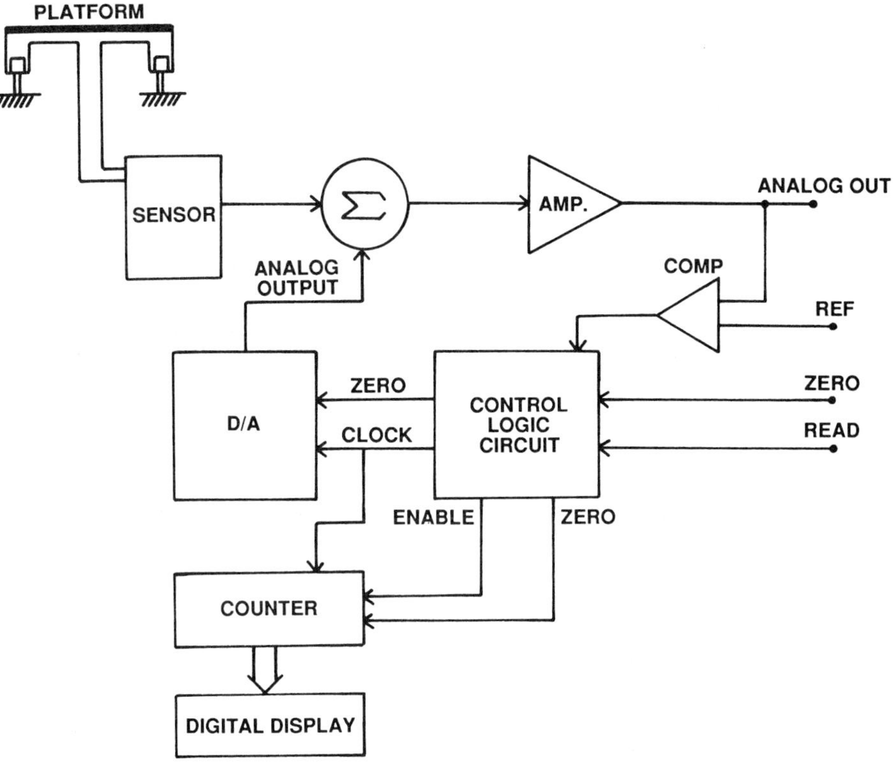


FIGURE 9.12   Digital scale.

and displayed. The cuff is then deflated to the reference condition
for the next measurement. The auto-zero ensures that each test
is referenced to zero pressure taking out all temperature- and time-
related offset errors. The price of the sensor can be reduced significantly.

*Digital Scale*

By adding another counter and display driver, the basic circuit
can be made into a scale capable of measuring tare weight. Figure
9.12 shows the circuit for a digital scale. A pressure sensor is
converted into a load cell by hydraulically coupling the sensor to
a platform. A 12- or 14-bit D/A converter is used to get adequate
resolution. With nothing on the platform, a zero command is given
to the auto zero, and the weight of the oil and platform are zeroed
out. If an object is placed on the scale and the read button is

pushed, the number of pulses required to rezero the circuit is
counted by the second counter and displayed. The process can
be continued until the full-scale pressure of the sensor is reached.
When the objects are removed from the scale, the zero button is
pressed, resetting all the counters to zero and rezeroing the empty
scale.

### Mixing Scale

Another variation of the basic circuit is shown in Figure 9.13.
If the reference voltage is changed, the same circuit could control
a mixing process. If the range of reference voltage is 0 to 10 V,
the circuit could be calibrated for 0 to 100 percent in 1 percent
steps.

### Automatic Zero Circuit

In some circuits, the output should never be negative or less than
zero; if the output is less than zero, it has to be a temperature
error or sensor error. Taking out the inhibiting circuit in the
standard auto zero gives a circuit that will automatically zero any
negative output voltage.

### Tracking Auto Zero

Another variation to the standard circuit is to make a tracking
sample and hold. In the standard circuit, the counters are reset
to zero at the beginning of each zero cycle. The counter then counts
pulses until the offset has been zeroed. If there is no error, the
counters would count to half scale, or an 8-bit DAC will count
128 counts. This can be a very time-consuming process if a 10-
or 12-bit DAC is used. The time to zero the circuit can be reduced
by converting the counters into up/down counters. The DAC is
not reset before each zero cycle; instead, a second comparator
is used to sense if the output is high or low and control whether
the counters count up or down (Figure 9.14). In this circuit, just
the change in error voltage is corrected each cycle by increasing
or decreasing the DAC output voltage. The first auto-zero cycle
after turn-on zeroes the system, just as the conventional auto zero
does.

### 9.3.5 Conclusion

The auto-zero technique can be used in many applications that
are cyclic and that have a reference pressure present at some time
in the sequence. A few inexpensive components can improve the

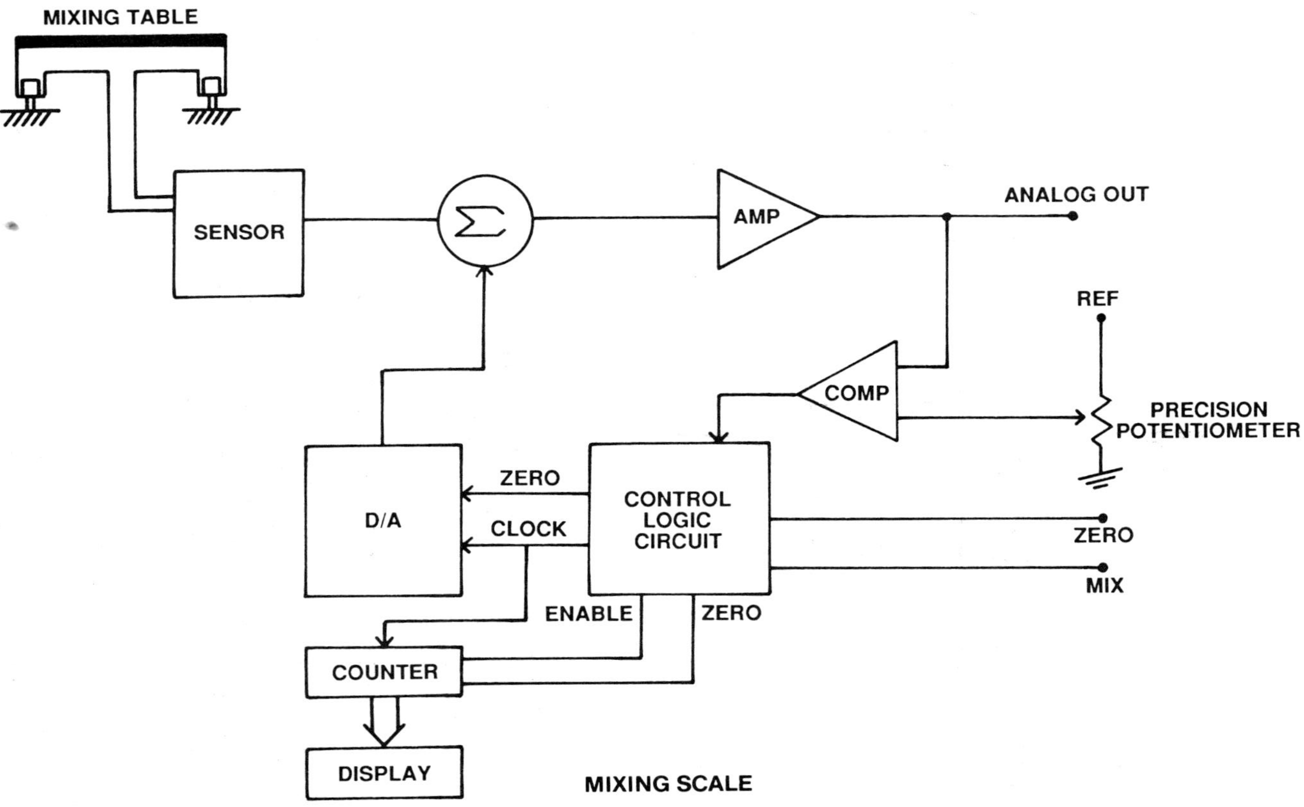


FIGURE 9.13  Mixing scale.

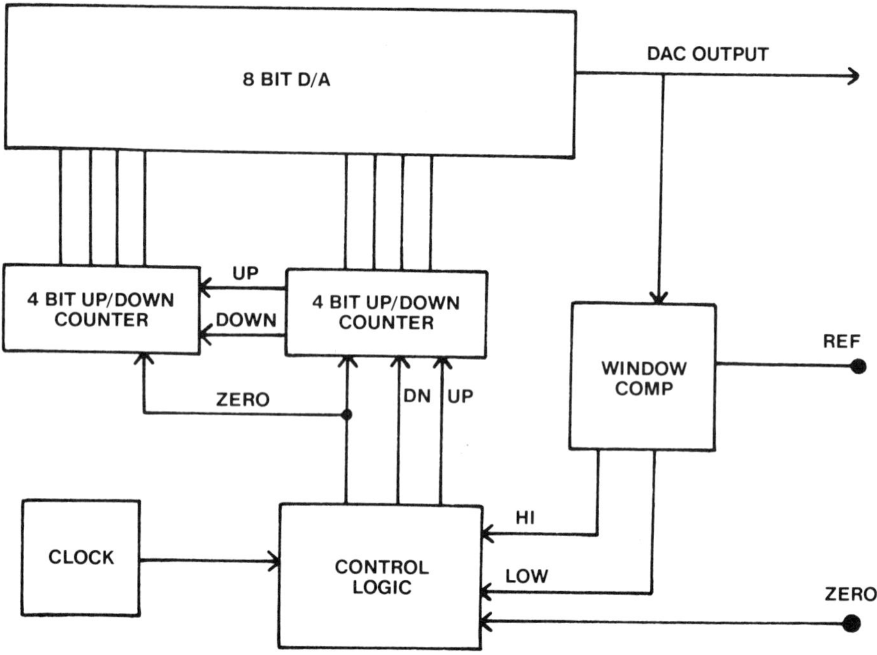



FIGURE 9.14   Tracking auto zero.

performance of the sensor by eliminating the offset errors. Using
a digital-to-analog converter for a sample and hold provides a sample
and hold with infinite hold time, good long-term stability, and controlled resolution (number of bits).

## 9.4   AUTO-REFERENCE TECHNIQUES

Using the memory and computing powers of the microprocessor
results in a very powerful tool to signal-condition sensors. Not
only the offset, but also the sensitivity, linearity, and pressure
range can be modified by the microprocessor.

The sensor can be compensated using the same basic technique
as used in the auto-zero method, by measuring the output of the
sensor at a reference pressure and storing this reference number
in memory. As each measurement is made, the offset error is electronically compensated by the microprocessor. The end result is
the same as that of the auto-zero circuit. The correction factor
is updated each time the reference pressure is applied to the sensor.

The auto reference can be taken one step further and a second reference pressure applied to the sensor to get a sensitivity measurement correction factor. This correction factor, as well as the offset, is stored in memory. The correction factors are updated each time the reference pressures are applied, to ensure that temperature and stability errors are corrected.

A third technique is to fully characterize each sensor and store all the data in memory instead of updating the correction factors periodically. A temperature measurement, as well as the sensor output voltage, needs to be taken at each measurement cycle. This technique is the most versatile of the three and is generally used for nonlinear applications such as flow and altitude measurements. Each sensor is characterized and the data are stored in the microprocessor memory. To make a measurement, the microprocessor measures the sensor output as well as the output of a temperature sensor. The microprocessor then looks up the appropriate correction factors for that particular temperature and modifies the sensor output by the magnitude of the stored data. Using this technique, the correction factors are not updated after the original sensor characterization. The technique requires a sensor with good stability specifications.


9.4.1  Basic Auto Reference

As previously described, the basic auto-reference technique does not require any hardware: A program is written that measures the sensor output voltage and stores the data in memory. Some care needs to be taken if an uncompensated pressure sensor is to be used. If the sensor has a large temperature error, a course-temperature compensation may need to be done to the sensor to ensure that the sensor and amplifier are not allowed to saturate due to temperature errors. The same converter (A/D converter, voltage-to-frequency converter) is used to measure the offset reading and pressure readings, which automatically compensates for any offset errors in the converter as well as in the sensor and amplifier.

The desired system accuracy will determine how often the stored reference number needs to be updated. A system that does not require high accuracy and works for short periods of time can auto-reference at turn-on and use that number until the power is turned off and back on.

A second approach is to update the reference number each time the reference pressure is present at the sensor. This technique works well in a cyclic application such as a blood-pressure measurement. The microprocessor is programmed to update the correction factor each time the cuff is vented.

A third approach is to have the microprocessor force a reference condition at the sensor by controlling a valve that disconnects the sensor from the system and connects it to a reference pressure (generally atmosphere) just before the auto-reference cycle. This approach requires some hardware—valve drivers and solenoid valves that must interface to the microprocessor. This method can be used in systems that do not have a reference condition, such as zero pressure or atmospheric pressure. This type of technique is generally used with low-pressure gas systems.

The reference pressure does not have to be zero. As with the auto-zero circuit, the auto-reference technique can be used for applications such as scales and pressure meters. The auto-reference technique has one advantage over the auto-zero technique: Readings can be stored or multiplied by constants to change the units displayed. If a display is used in the system, the microprocessor can be programmed to automatically change the display range (like an auto-ranging digital voltmeter) to get maximum resolution from the measurement.


## 9.4.2  Characterizing the Sensor

Since each sensor is a little different in temperature characteristics, each sensor has to be measured accurately over the desired temperature range. This technique would be very expensive if a small number of units were to be compensated, but an automatic system could measure hundreds of units and automatically blow a PROM for each device. The system could also be used to check the stability and repeatability of each device before the PROM is blown. Sensor characteristics that can be measured and modified include

1.  Offset vs. temperature
2.  Sensitivity vs. temperature
3.  Linearity
4.  Repeatability

The amount of memory available will determine how many data points can be used to compensate the sensor. The more data points used, the more accurate the compensation. Factors that determine how much data must be stored are

1.  PROM size
2.  Compensated temperature range
3.  Pressure range
4.  Compensated accuracy desired

If the temperature characteristics are linear over the desired
temperature range, an equation can be derived and stored in memory.
If an equation cannot be derived, a lookup table must be used.

How the sensor is tested is also important to get accurate
measurements. The sensor should be tested as closely as possible
to the configuration used in the final product, at the same voltage
or current that will be used in the final product. Also, the sensor
should be tested in the final package configuration, if possible,
to ensure that the sensor will be exposed to the same mechanical
stresses as the finished product.

At each temperature, there will be an offset correction (independent of pressure) and a sensitivity correction that is a function
of applied pressure. The sensor is connected to a pressure source
and measured at a number of pressures to get linearity data for
each temperature. The linearity data can be taken over an extended
pressure range to maximize sensor sensitivity; linearity will be
pressure-dependent. In high-volume applications, the user and
sensor manufacturer should work together to select the optimum
chip for the particular application. Diaphragm thickness and chip
design can be selected to optimize the desired sensor characteristics
for the application.

## 9.5  AUTO-REFERENCE APPLICATIONS

Just about any system that uses a microprocessor and a pressure
sensor can use some form of auto reference. The most common auto
reference is the zero correction. It is relatively easy to program
and requires a minimum of microprocessor memory.

### 9.5.1  Offset Auto Reference

As previously discussed, automatic blood-pressure-measuring systems
and electronic scales are well suited for auto-reference zeroing,
since they are both cyclic in nature and return to a reference condition after each measurement. There are some applications that
are not as straightforward, but can also use a form of auto reference.

#### *Liquid Level Measurements*

Some liquids are stored in containers that are pressurized by the
evaporating liquid—the liquid evaporates until a head pressure is
established equal to the evaporation pressure of the liquid. The
head pressure may be regulated and vented to atmosphere or sealed.
To measure the volume of liquid in the container, the head pressure

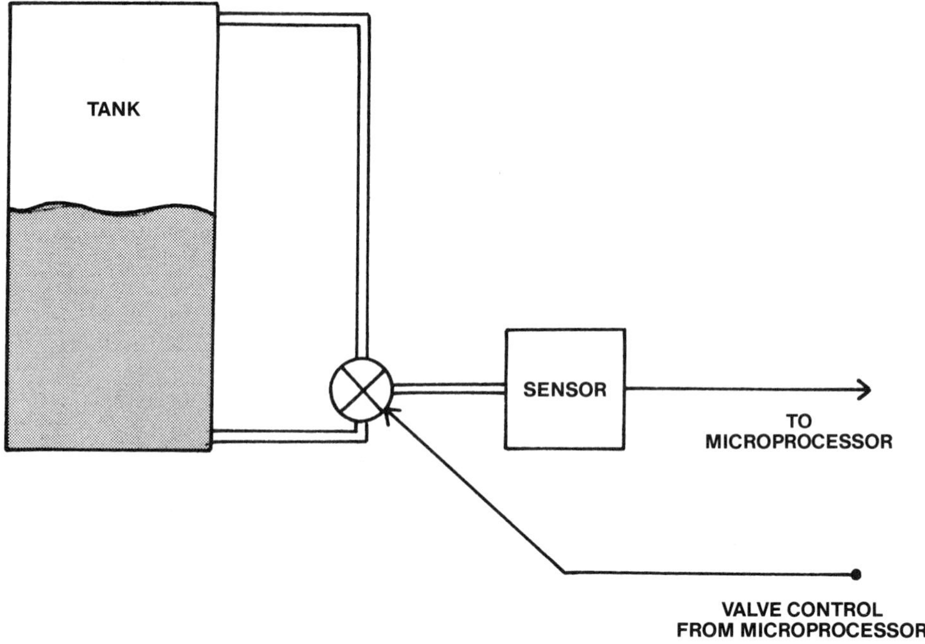


FIGURE 9.15  Liquid-level measurements.

and the pressure at the bottom of the container must be measured
to calculate the liquid level. As shown in Figure 9.15, one gage
sensor can be used to measure both the head pressure and the
pressure at the bottom of the container by installing a two-way
valve that the microprocessor switches from tap to tap. Using the
same sensor to measure both pressures eliminates any offset errors
in the system. The microprocessor can also be used to monitor
the head pressure to ensure that it doesn't increase to an unsafe
level, as well as monitoring the rate of use and giving some indication of when it is time to reorder. This is a relatively simple application: Two measurements are taken and subtracted to get the
differential pressure.

*Barometric Measurements*

In some applications, changes in pressure are more important that
the absolute value. Barometers are normalized to sea level, and
the magnitude and rate of change are used to indicate a weather
front passing through. An electronic barometer is an ideal application
for auto-reference techniques. A reference number is taken and

stored; the barometric pressure can be obtained from a local airport or weather station and hand-programmed into the system for a reference point.

The system can be programmed to perform a number of different functions. The simplest system would measure the absolute pressure and display it. A more complex system might give time-related readings and indicate changes in the barometric pressure over a period of time. Since barometric pressure normally changes very slowly, the system could be programmed to take a reading every 60 minutes or so and store the pressure readings and the time measured. The system also could calculate the rate of change as a function of time and indicate if the pressure is rising or falling and how fast. The system could be programmed through a terminal or keyboard to display the desired data.

### 9.5.2  Electronically Compensated Sensors

Electronically compensated sensors have made a number of high-accuracy applications possible.

*Smart Transmitters*

The addition of a microprocessor and characterized sensors to industrial pressure transmitters has turned the transmitter into a small system. The new smart transmitters have features such as self-diagnostics, digital programmability in the field, and over-range measurement capability. Comparing a standard 4 to 20-mA transmitter with the smart transmitters:

| Specification | Standard | Smart |
| --- | --- | --- |
| Output | 4 to 20 mA | 4 to 20 mA |
| Accuracy | ±0.5% | ±0.1% |
| Digital damping | N/A | 0 to 32 s |
| Combined zero and span TC | ±1% | 0.5% |
| Turndown | 4:1 | 16:1 |

The biggest improvements are in accuracy and turndown. The microprocessor is used to compensate for normal linearity errors, which increases the accuracy of the transmitter. The other significant improvement is in the transmitter's turndown range. Since one transmitter can be used to measure more pressure ranges, less ranges have to be stocked to measure the same pressure range.

The smart transmitter also has the advantage that the output of
the microprocessor can be buffered and used as a digital output
to interface with a central computer. The key to the smart transmitter is that each sensor is characterized and the microprocessor
is used to compensate each sensor.

*Altimeters*

Electronic compensation works well for nonlinear applications such
as altitude measurements and flow measurements. The sensor is
characterized over the desired temperature range, just as is a
standard sensor. The appropriate altitude, instead of the linear
pressure values, is stored in memory. As each pressure reading
is taken, the microprocessor compensates for thermal errors and
converts the linear pressure reading into a nonlinear altitude reading.
The altitude-vs.-pressure relationship is shown below:

| Altitude (ft) | Pressure (in.Hg) |
|---:|---:|
| 0 | 29.9213 |
| 500 | 29.3846 |
| 1,000 | 28.8557 |
| 1,500 | 28.3345 |
| 2,000 | 27.8210 |
| 3,000 | 26.8167 |
| 4,000 | 25,8418 |
| 6,000 | 23,9782 |
| 8,000 | 22.2250 |
| 10,000 | 20.5770 |
| 12,000 | 19.0294 |
| 14,000 | 17.5774 |
| 16,000 | 16.2164 |
| 18,000 | 14.9421 |
| 20,000 | 13.7501 |
| 25,000 | 11.1035 |
| 30,000 | 8.8854 |
| 35,000 | 7.0406 |

The chart shows the relationship between altitude and pressure.
From sea level to 1000 ft, the pressure changes 1.0656 in.Hg/1000
ft, and at 20,000 ft, the pressure changes 0.596 in.Hg/1000 ft.

*Backpacker's Altimeter*

Using the same concept, a low-altitude altimeter (say, 0 to 12,000 ft)
could be designed that would also work as a barometer and thermometer.

The microprocessor has to measure temperature to compensate the sensor, and it would be straightforward to program the microprocessor to display the temperature on command. Also, at a constant altitude, the microprocessor could be programmed to display the change in barometric pressure over the last few hours to indicate a weather front.

*Flow Meters*

Measuring a pressure change across a venturi or calibrated restriction is a popular flow-measuring technique. Electronically compensating a differential sensor gives a versatile compensation. Parameters such as common-mode pressure sensitivity can be compensated at the same time. The three most common problems encountered in using this technique are that the pressure drop across a fixed restriction is nonlinear; at low flows, the pressure drop is very small; and very small pressure changes have to be measured at high working or common-mode pressures. Flow meters may have to measure decreases of a few inches of water pressure at working pressures up to 1000 pounds. The high working pressure can cause an additional error in the sensor offset voltage, due to the package being stressed. The pressure-vs.-temperature compensation is performed as before. A second set of measurements recording offset vs. working pressure could also be stored, and a second pressure sensor could be installed in the system to measure working pressure. The large dynamic range of the measurements could be handled by using two differential sensors (high and low pressure).

## 9.6  CONCLUSION

A microprocessor-based system is a very powerful tool, for several reasons. By adding some software, the system can be auto-referenced, increasing the system accuracy. The offset errors can be compensated with a simple auto-reference technique, or the sensor and associated electronics can be electronically compensated by characterizing the pressure sensor and storing the correction factors in the microprocessor memory. Accurate systems can be designed using sensors that have been selected for repeatability and stability.

# 10

# Computer Interface Techniques

## 10.1 INTRODUCTION

As more and more control and monitoring systems become computerized, there is an increased need to convert the analog output of sensors to a digitized format that can interface to the computer. The analog output needs to be converted to a frequency, a variable pulse, or digital pulses. Microprocessors are being designed into products ranging from cameras to cars, and each application is unique. The optimum technique for a particular application will depend on factors such as the number of sensors in the system, the system accuracy, the speed of the system, and—always—cost.

Some of the most common digitizing techniques include

1. Parallel A/D converter
2. Serial A/D converter
3. Variable frequency
4. Variable pulse width
5. Variable duty cycle

The digital input can be put directly on the microprocessor bus if the converter is microprocessor-compatible and a parallel format is used, or it can be read through a serial port on the microprocessor if a serial format, such as frequency, is used. The microprocessor must be able to address the converter and give read and write commands to control the flow of information. In some applications, a number of sensors are fed into a multiplexer, and the microprocessor controls the data from many sensors by selecting one sensor at a time and storing the data. Data-acquisition systems may have many channels of data fed into one microprocessor or computer.

## 10.2  COMPUTER INTERFACE FUNDAMENTALS

### 10.2.1  Definitions

To get a better understanding of the process of interfacing sensors and converters to computers, some of the terms commonly used will be defined in this section.

*Accuracy*:   Total accuracy of converters consists of the sum of nonlinearity error, zero error, gain error, and reference error specifications.

*Aperture*:   The period of time that the input to an A/D converter should not change more than 1/2 LSB. The aperture determines the maximum sampling frequency of the A/D converter.

*Bit*:   A digital signal that has two states, high or low.

*Bus*:   A parallel path of binary information signals—usually 4, 8, or 16 bits wide.

*Byte*:   A binary digital word, generally 8 bits long.

*Chip Select*:   A logic input signal that activates data or control information transfer between the computer and converter. This signal is designated by CS. Chip select is used in conjunction with other logic signals to determine the exact activity to take place.

*Clock*:   All A/D converters require a clock. The clock sets the basic conversion rate of the converter. The clock may also be driven by the system or computer clock for timing.

*Coding*:   The format of the A/D output data. Common formats include binary, offset binary, twos complement, low byte, and high byte.

*Control Lines*:   Digital input/output pins that activate, monitor, and control A/D operations; for example, chip select, write, low byte, high byte, start convert, conversion complete, busy, read, etc.

*Conversion Complete (or Busy)*:   The digital output signal that shows the status of the converter—one state busy or converting state, the other state conversion complete.

*High Byte (HB)*:   A/D converters with more than 8 bits may offer a 2-byte format to simplify interfacing to 8-bit systems. The

high byte contains the most significant bit and some or all of
the upper 8 bits.

*Left-Justified Data*:  In the byte-oriented data output format, data
bit sets shorter than 8 bits are placed starting in the left side
of the data output transfer register. This could apply to the
upper or lower byte.

*Least Significant Bit (LSB)*:  The binary digit with the smallest
numerical weighting. Normally, one LSB equals the full-scale
range divided by $2^n$ ($n$ is the number of bits in the A/D).

*Low Byte (LB)*:  A/D converters with more than 8 bits may offer
2-byte operation to simplify interfacing to 8-bit systems. The
low byte contains the least significant bit and some or all of
the low 8 bits of the A/D output.

*Quantization Error*:  An A/D converter can identify only $2^n$ output
codes. Each output code represents a range of analog voltages.
The ±1/2 LSB limit to resolution is known as the *fundamental
quantization error*.

*Reference Voltage* ($V_{REF}$):  A precise voltate that determines the
full-scale output of sensors and converters. The stability and
temperature characteristics determine the sensitivity characteristics of the sensor or converter. In some systems, the same
reference voltage is used for both sensors and converters.
If the reference is common to both and the effect of a changing
reference is 180 degrees out of phase, the reference errors
cancel (converter—increase in reference voltage decreases sensitivity, sensor—increase in reference voltage increases sensitivity).

*Resolution*:  The minimum change the device can detect. The resolution of an A/D converter is the number of states into which
the input voltage is divided ($2^n$), where $n$ is the number of
bits. The resolution of a sensor is determined by the signal-to-
noise ratio of the sensor. Resolution should not be confused
with accuracy.

*Right-Justified Data*:  In the byte-oriented data output format,
data bit sets are placed starting on the right side of the data
output transfer register. This could apply to the upper or
lower byte.

*Sampling Frequency*:  The rate at which an A/D converter can
continuously convert analog inputs. The sampling frequency
is determined by the clock frequency and the type of conversion.

*Span*:  Used to describe the full-scale analog input voltage of converter products or the change in the sensor output voltage
when the input pressure is changed from minimum to maximum
specified pressure.

*Successive-Approximation Conversion*:   Successive-approximation
    converters compare an unknown analog input voltage against
    an accurately known binary fraction of the reference voltage.
    Starting with the most significant bit (1/2 reference voltage),
    the converter will determine if the unknown is greater or less
    than the known voltage. If the unknown is less that the com-
    pared voltage, the bit is turned off and the next smaller bit
    is turned on. This process continues until the least significant
    bit has been sampled and the status of each bit is stored in
    the SAR (successive approximation register).
*Transition Noise*:   Width of the transition voltage that is undefined.
*Transition Voltage*:   The center of a band of input voltages where
    the digital output code toggles between two adjacent codes.
*Tri-State Output Buffer*:   A digital output stage that has a mode-
    control input pin. In one mode, the output is active and capable
    of driving the data bus; in the other, the device is in a high-
    impedance mode and will not affect the state of the data bus.
    The device is left in the high-impedance mode when other devices
    are controlling the bus.
*Word*:   A set of binary digits that represent the fundamental register
    size of the digital circuits being used; for example, 4-bit, 8-bit,
    or 16-bit systems.


## 10.2.2  Interface Basics

In a computerized measuring system, the microprocessor or computer
controls the step-by-step operation of the system. Some of the
basic functions the microprocessor controls are as follows:

1.  System functions
    a. Turn relays and valves on and off
    b. Control multiplexers
    c. Select the sensor to read
    d. Control converters
    e. Interface with other computers in system
2.  Converter functions
    a. Start conversion
    b. Wait until EOC (end of conversion) changes state
    c. Read A/D output register—or, if the A/D is more than 8 bits:
       (1) Read high byte
       (2) Store high byte
       (3) Read low byte
       (4) Store low byte
    d. Loop back to start

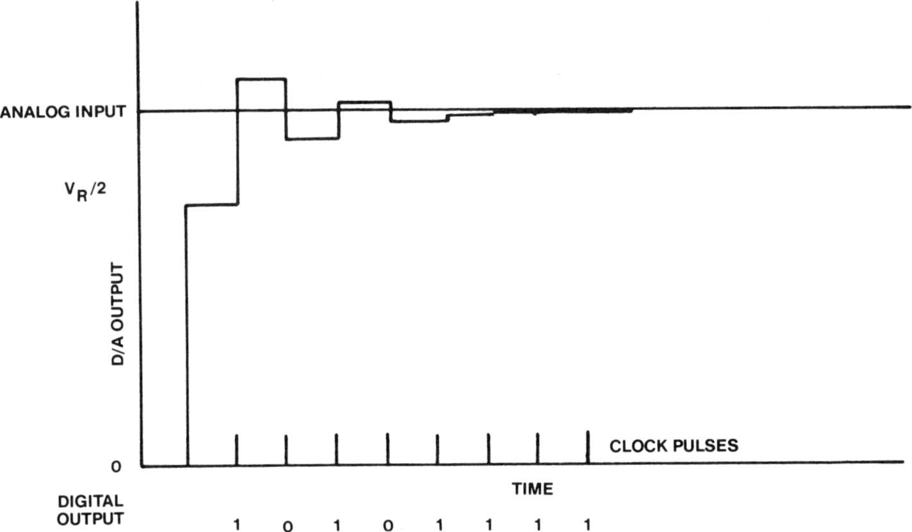



FIGURE 10.1   Successive-approximation sequence.

To perform functions other than monitoring, a computer must
be able to control some physical parameters, such as controlling
a flow by turning valves on and off, or turning motors on and
off by controlling relays. To do this, special interface driver cards
have to be purchased or designed to interface the computer to the
outside world.

The microprocessor has to be able to communicate with the
converter to give commands to start conversion and read output.
If the converter is not microprocessor-compatible, interface devices
must be used (Figure 10.1) to interface the microprocessor to the
sensor. If the converter is to interface to the system bus, but
the converter has a standard transistor transistor logic (TTL) output, a tri-state buffer must be used between the A/D and the bus,
and additional logic must be added to control the chip-select function
as well as the read cycle. This is also true if the sensor has a
serial interface—the computer has to be aware when the converter
is sending information and be able to read the converter output.
When more than one sensor is in the system, the microprocessor
must be able to distinguish one converter from another and read
each output.

## 10.3  PARALLEL ANALOG-TO-DIGITAL CONVERTERS

There are many types of parallel analog-to-digital converters, and
each has advantages and disadvantages. Some of the primary features
of each conversion technique are shown in the following chart:

| Type | Speed | Resolution |
|------|-------|------------|
| 1. Successive approximation | Medium ($\mu$s) | Medium-high (8-16 bits) |
| 2. Flash | Fast ($\mu$s) | Low-medium (4-8 bits) |
| 3. Single slope | Slow (ms) | Medium-high (8-14 bits) |
| 4. Dual slope | Slow (ms) | High (10-18 bits) |
| 5. Residue multiplication | Slow (ms) | High (10-18 bits) |

Parallel output converters have one output line per bit, and
the state of all the bits is read at one time. Parallel output formats
are useful if the system is compact and the signals do not have
to be transmitted over long distances. The two most popular conversion techniques are successive approximation and dual slope.
Successive approximation is high speed and available with 8-, 10-,
and 12-bit resolution. Dual slope is much slower, but is available
with 14- to 18-bit resolution (high accuracy).


### 10.3.1  Successive Approximation

The successive-approximation A/D converter (ADC) compares the
unknown input to each bit, starting with the most significant bit.
The converter turns on the most significant bit and compares it
to the analog input voltage. If the input voltage is greater than
the first bit, that bit is left on and the best bit is turned on.
If the voltage is less than the most significant bit, the bit is turned
off and the next bit is turned on. This process continues until
the unknown voltage has been compared to all the bits; the status
of each bit is then stored in the logic register (Figure 10.1). The
status of each bit is then loaded into an output buffer to be transferred to the system on command.
   The successive-approximation analog to digital converter
(Figure 10.2) consists of a digital-to-analog converter, a clock,
a reference voltage, an output buffer, and an SAR. The SAR

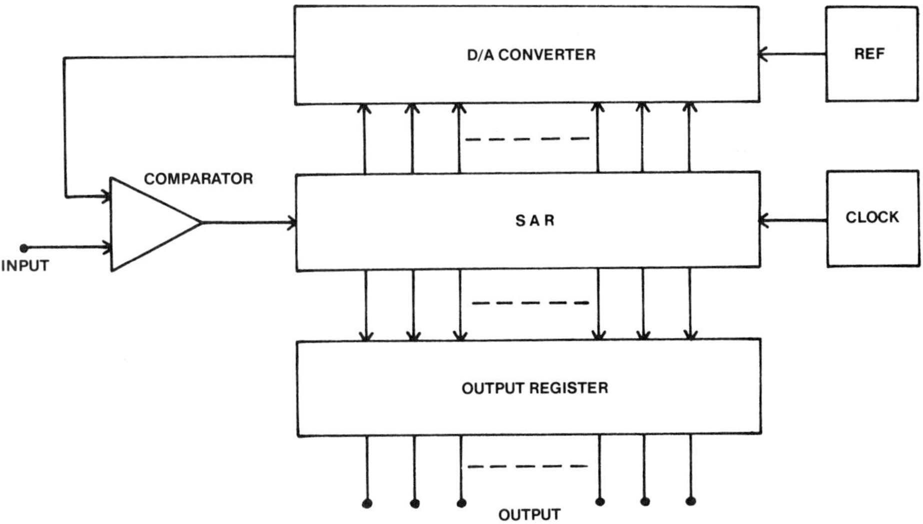



FIGURE 10.2   Successive-approximation A/D converter.

contains all the logic necessary to turn on each bit and decide
to leave it on or turn it off, depending on the input from a comparator that compares the DAC output to the unknown input voltage.
The reference-voltage accuracy determines the final specifications
of the ADC; if the reference voltage moves, the converter gain
will be off by the same percentage. The converter performs the
basic conversion in 8 to 12 clock pulses, depending on the number
of bits in the ADC. A 10-bit successive-approximation ADC could
do a conversion in 5 to 10 μs, depending on the clock frequency.
    Some of the new A/D converters have all the logic necessary
to interface to a microprocessor; they are generally called
microprocessor-compatible ADCs. Figure 10.3 shows a 10-bit
microprocessor-compatible A/D converter. The 10-bit DAC is controlled by the SAR; the output of the DAC is compared to the
unknown input by the comparator and the output is fed back to
the SAR. The output of the SAR is also fed to a tri-state buffer.
(A tri-state buffer can be programmed to output a logic low, logic
high, or high-output impedance state. This device can be connected
directly to the digital bus. When it is not driving the bus, it is
left in the high-impedance state, which isolates the ADC from the
bus.) The other logic inputs that have been added are chip-select
inputs. The CS inputs allow the microprocessor to address the
A/D converter and control the function it performs.

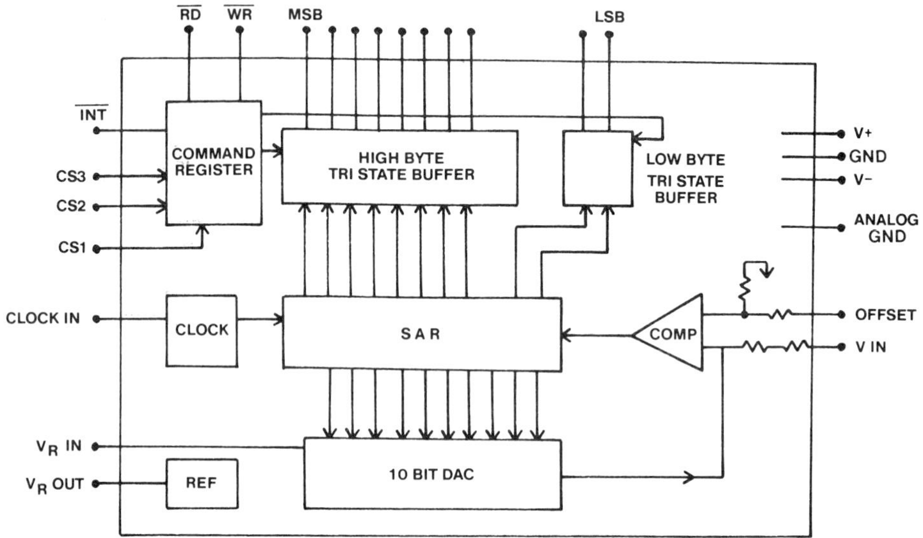

FIGURE 10.3   Microprocessor-compatible A/D converter.


The other function added to the ADC is the command register, which can be programmed by the user to select various modes of operation, such as unipolar or bipolar operation, range selection, or output data format. All these inputs are controlled by the data bus. The normal start and busy are used; unfortunately, there is no set logic format used with microprocessors, so some of the commends may have to be inverted for use with certain microprocessors.


## 10.3.2   Flash Converters

The flash converter gets its name from its ability to do analog-to-digital conversions very rapidly. The flash converter does a conversion every time it receives a clock pulse.

The flash-conversion speed is accomplished by providing a comparator for every quantization level except one (either zero or full scale). By using $2^{n-1}$ comparators, all possible combinations are examined simultaneously, and the conversion is accomplished in one clock period. The limiting factor of this method is the number of comparators required. For example, a 4-bit ADC requires 15 comparators, an 8-bit ADC requires 255 comparators, and a 12-bit ADC requires 4095 comparators. The practical limit for the number of comparators is 8 or 9 bits.

As shown in Figure 10.4, the unknown input voltage is connected to one input of all the comparators, and the other comparator input is connected to a resistor ladder network that divides the reference voltage into $2^n$ equal voltages. At the rising edge of the clock pulse, all comparators that are at a reference voltage greater than the unknown voltage are latched 0 or low state, while all comparators with a reference voltage less than the unknown voltage are latched 1 or high state. A decoder network converts the comparator outputs into an $n$-bit digital word that is stored in an output buffer.

### 10.3.3  Single-Slope Analog-to-Digital Converter

The single-slope integrator arrives at its digital output by comparing the unknown analog input to a ramp generated from a reference voltage. The ramp must be precisely controlled so that it is very linear and takes exactly $2^n$ clock pulses to go from 0 to full scale. By counting the number of pulses required to build a ramp from 0 V to the value of the unknown input voltage, the digital output is obtained.

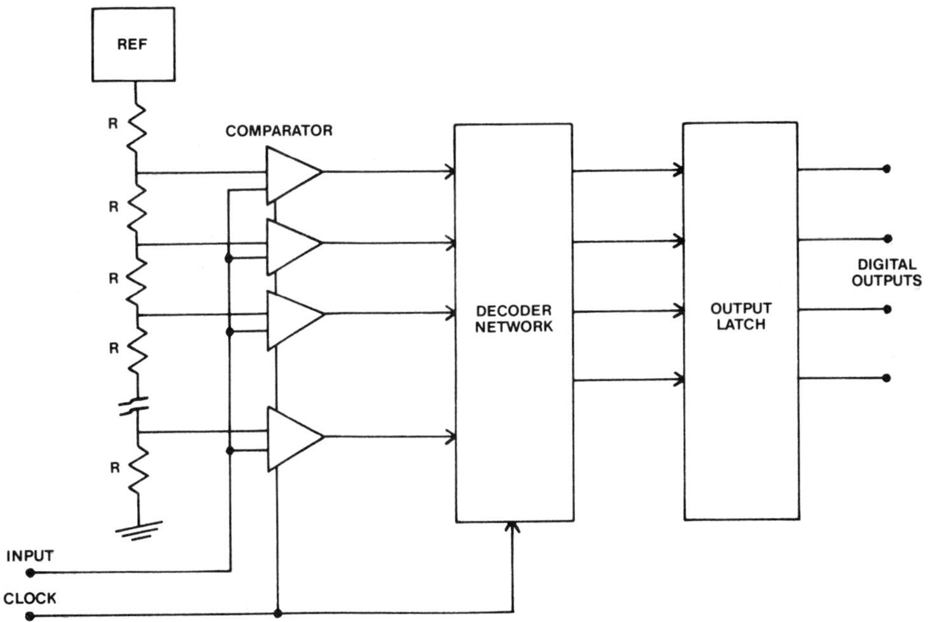


FIGURE 10.4  Flash converter.

### 10.3.4  Dual-Slope Analog-to-Digital Converter

Most high-accuracy applications, such as 5- and 6-digit voltmeters,
use some variation of dual-slope conversion. Dual-slope conversion
has some built-in advantages in that noise is rejected when the
input voltage is integrated, and very high accuracies can be achieved
(10 to 18 bits). Dual-slope conversion is an indirect method for
A/D conversion where the unknown analog voltage and a reference
voltage are converted into time periods by an integrator and counted
by a counter. The dual-slope integrator arrives at its digital output
by building ramps. The first integration (ramp) is accomplished
by using the unknown input voltage and integrating at a rate proportional to the input voltage for a fixed period of time. The peak
value of this integration is proportional to the unknown input voltage.
The second ramp is generated when the capacitor is discharged at
a precise rate determined by the reference voltage. The discharge
time is measured precisely and converted into the digital output
(Figure 10.5). Using the same clock to count time ($T$) and ($t$),
and using the same integrator network to generate both ramps,
allows the clock-frequency errors and the integrator errors to cancel.
This leaves the stability of the reference voltage as the determining
factor in the overall accuracy. The block diagram of a dual-slope
A/D converter is shown in Figure 10.6.

### 10.3.5  Modified Dual-Slope Technique

Some of the newer, high accuracy converters use a version of dual-
slope conversion called *successive integration*. These converters
enhance the dual-slope conversion technique through multiple dual-
slope conversions, with each successive conversion having 10 times
the resolution of the previous conversion. The key to this successive-
integration technique is the multiplication of the residual voltage
on the integration capacitor after each conversion. As shown in
Figure 10.7, at the end of the first deintegrate cycle, the cycle
terminates one clock pulse after the integrator output crosses zero,
leaving a small residue voltage on the integration capacitor. This
residue voltage is then multiplied by 10 and a second dual-slope
conversion is performed. The second conversion will have 10 times
the resolution of the first conversion. The second integration cycle
increases the resolution of the converter by a factor of 10, or one
digit. The residue multiplication process can be repeated a second
time, increasing the resolution of the converter by one more decade
or one more digit. This technique can improve the resolution of a
dual-slope ADC by a factor of 10 or more.

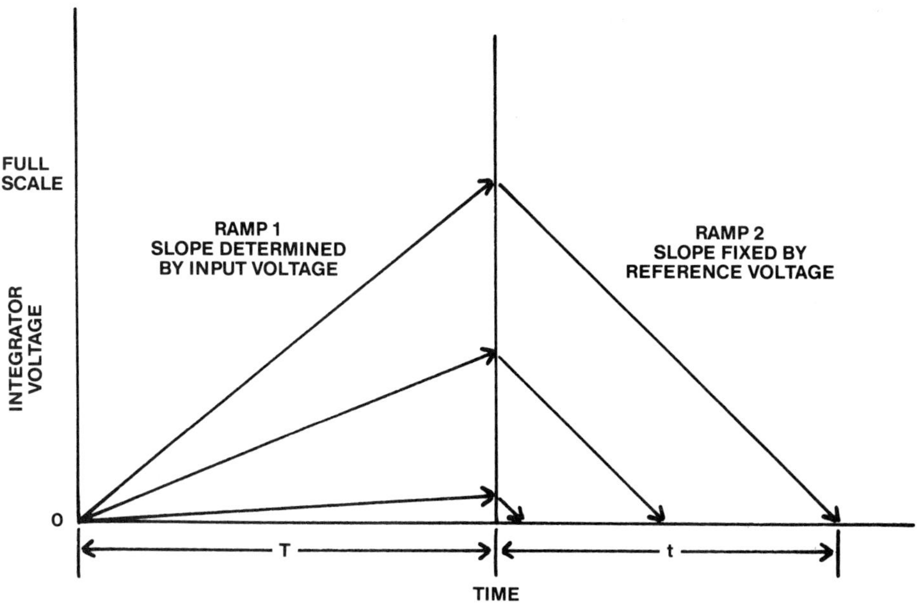


FIGURE 10.5   Dual-slope sequence.

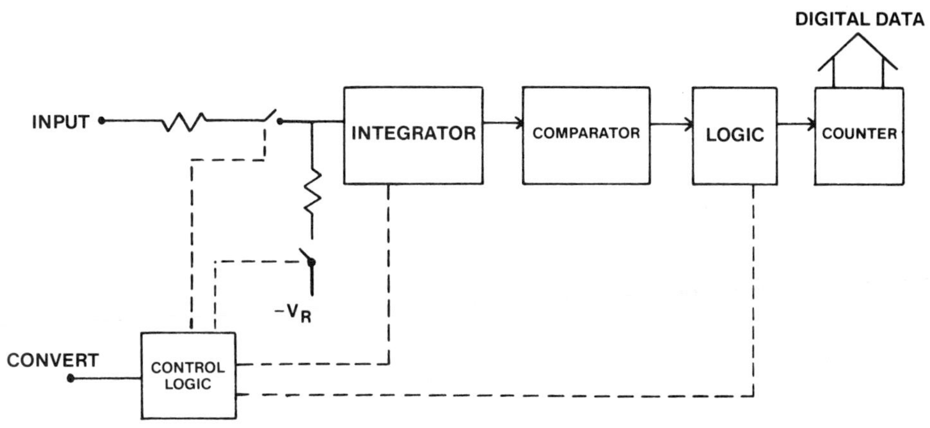


FIGURE 10.6   Dual-slope converter.

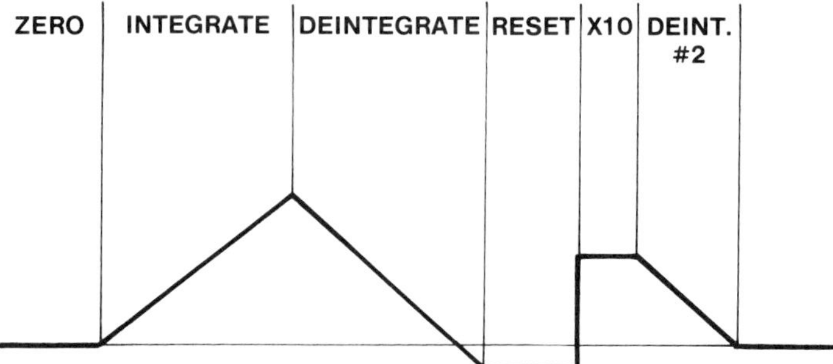


FIGURE 10.7   Integrator waveform of modified dual-slope converter.

## 10.4   SERIAL ANALOG-TO-DIGITAL CONVERTERS

To operate, parallel converters require one wire per bit, plus at least three control lines. The rising price of copper wire makes the parallel converter impractical when the sensors and microprocessor are separated by some distance. In these applications, some form of serial converter would be considered.

Serial converters can be two-wire or three-wire systems. Four- to twenty-milliamp transmitters are used in two-wire systems, where power and signal share the same two wires. Most serial converters are three-wire systems, two wires for power and a third for output signal (the common lead is shared by power and signal). Any device that converts an analog input voltage to a digital output (binary, pulse, frequency) and uses only one signal lead is considered a serial output device. Some common serial converters include

* Successive approximation (with serial port)
* Voltage-to-frequency converter
* Voltage-to-pulse-width converter

### 10.4.1   Successive-Approximation Converters

Any parallel analog-to-digital converter can be used in a serial mode by adding a parallel-to-serial converter. Converting from parallel to serial is accomplished by loading the digital output into a shift register and shifting the output out of the register one bit at a time. The microprocessor reconstructs the digital word by reversing the process—loading the bits into an input register as they are received and converting back to a parallel format. Any ADC can be converted to serial output, but successive-approximation converters are the most common.

The key to transmitting serial data is timing. The microprocessor has to know when to look at the input for each bit and when each bit starts and stops. The data are divided into $n + 2$ time periods, 1 start bit followed by $n$ data bits and 1 stop bit (Figure 10.8).

A successive-approximation converter does a standard conversion and then loads the data into an output register, where they are shifted one bit at a time to the serial port.

### 10.4.2 Voltage-to-Frequency Converters

Another way to transmit serial data is to drive a voltage-to-frequency (V-F) converter with a sensor output. As the name suggests, the voltage-to-frequency converter converts the analog sensor output to a linearly changing frequency. There are many ways to convert a voltage to a frequency; the best design for a particular application depends on the desired precision, speed, response time, dynamic range, and other considerations. Most V-F converters generate a ramp voltage by controlling the current that charges a timing capacitor. This ramp voltage is used to trip a comparator, which generates the output frequency of the oscillator. The technique used to charge and discharge the timing capacitor depends on the desired accuracy and frequency range of the V-F converter.

*Integrator V-F:*   Figure 10.9 shows a basic oscillator circuit, where an input voltage is converted to a proportional current by the input resistor. This current charges a capacitor in the feedback loop of an integrator, generating a voltage ramp at the output. When the ramp crosses $V_{REF}$, the comparator changes state, and the switch in parallel with the capacitor is closed, discharging the integrator capacitor and restarting the cycle. The output frequency is proportional to the input voltage. The drawback to this approach is the discharge time of the integrator capacitor,

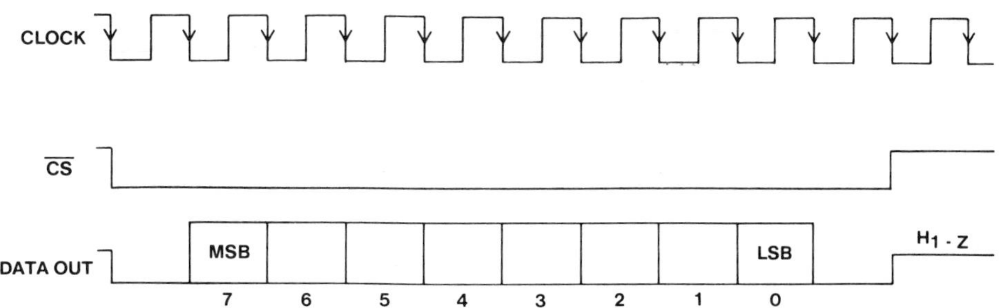


FIGURE 10.8   Serial-output timing diagram.

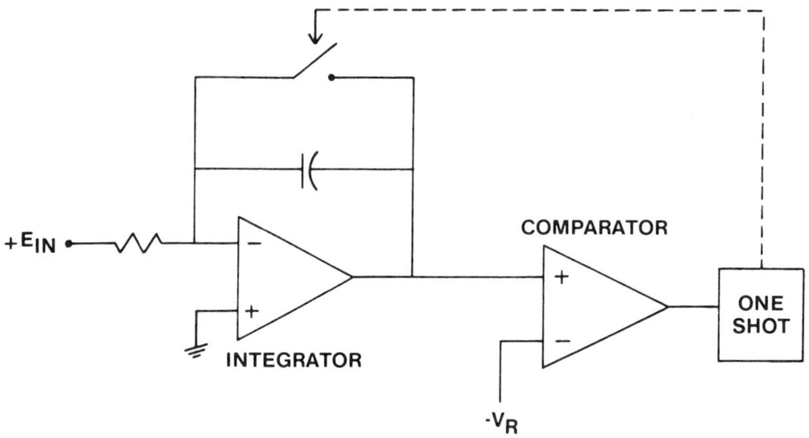


FIGURE 10.9   Integrator-comparator V-F.

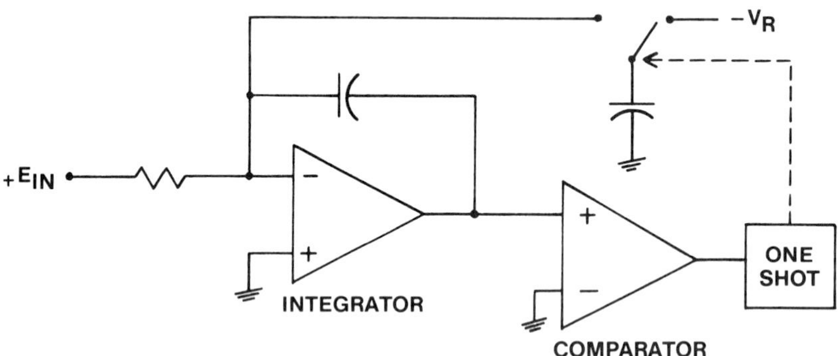


FIGURE 10.10   Charge pump V-F.

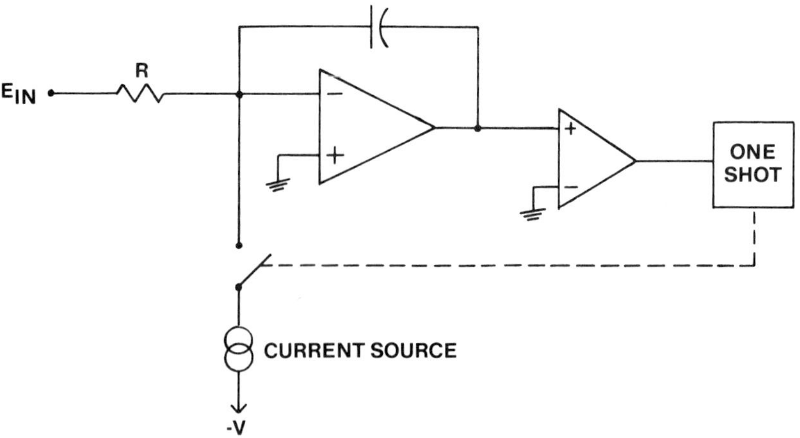


FIGURE 10.11   Current balance V-F.

which results in a significant linearity error at higher frequencies. Performance of the basic circuit is limited.

*Charge Pump V-F*:   The basic circuit can be modified to get around its limitations by charging a capacitor to a fixed reference voltage during the integrate time, as shown in Figure 10.9. When the comparator changes state, capacitor $C_1$ discharges into the integrator summing junction and forces the output of the integrator high, which decreases the reset time and improves V-F linearity (Figure 10.10).

*Current Balance V-F*:   The circuit shown in Figure 10.11 uses a current instead of a charge to maintain the operational-amplifier summing point. Each time the integrator output crosses zero volts, the comparator triggers the one-shot and connects the current source to the integrator's summing junction, forcing the output positive. At the end of the current sink period, the integrator output starts the next cycle ramping negative. The frequency of this action depends on the input voltage.

*Relaxation Oscillator V-F*:   The relaxation oscillator generates a ramp with a fixed slope and changes the frequency by changing the comparator reference voltage. As shown in Figure 10.12, an RC network is charged by the current source to a voltage higher than the input voltage. When the one-shot turns off, the switch is opened and the capacitor discharges through resistor $R_1$, generating a negative ramp until the input voltage is reached. When the ramp reaches the input voltage, the one-shot turns on, charging the capacitor for the next cycle.

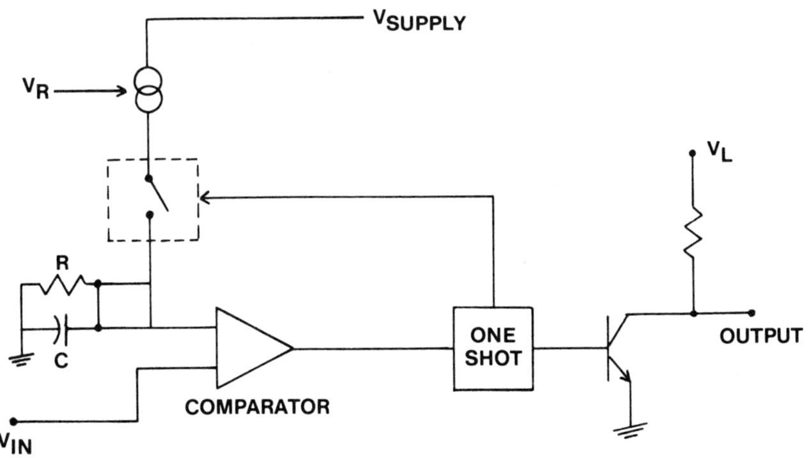


FIGURE 10.12   Relaxation oscillator V-F.

Voltage-to-frequency converters can be purchased in a number of technologies: monolithic ICs, hybrids, or functional building blocks. Each application generally requires a certain frequency range, package size, and price range. If a commercial V-F converter cannot be found that meets the requirements, one can be designed using function blocks such as switches and amplifiers.

To use a voltage-to-frequency converter with a pressure sensor, the output of the sensor is amplified and level-shifted by an amplifier stage to be compatible with the input of the V-F converter. The output of the V-F converter is connected to the microprocessor, or to a buffer or multiplexer if more than one sensor is to be connected to one port of the microprocessor.

The pressure-sensor output must be made compatible with the V-F converter by amplifying the sensor output and converting it from differential to single-ended. The pressure sensor and V-F converter make up a frequency-output transducer. Since the output is a changing frequency, these devices are less susceptible to induced noise voltage and line impedance than is an analog voltage transducer. These frequency-output devices can be transmitted over long distances and, in some instances, can share signal and power lines.

The V-F converter can be monitored by (1) connecting its output to the input of a digital counter and counting the pulses for a precise period of time, or (2) connecting the V-F output to the serial input of a microprocessor and programming the microprocessor to count the pulses for a precise time period. This stored digital word is proportional to the measured pressure at the sensor. The resolution is determined by how accurately the frequency can be monitored and the stability of the voltage reference in the voltage-to-frequency converter.

If a tri-state buffer is added to the output of each voltage-to-frequency converter, a number of converters can be connected to

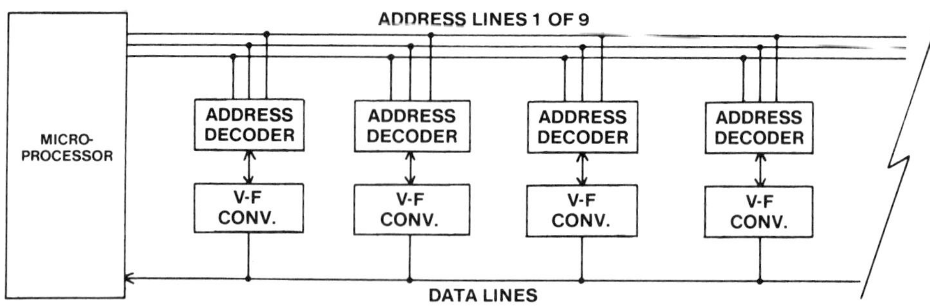


FIGURE 10.13  Time-share system.

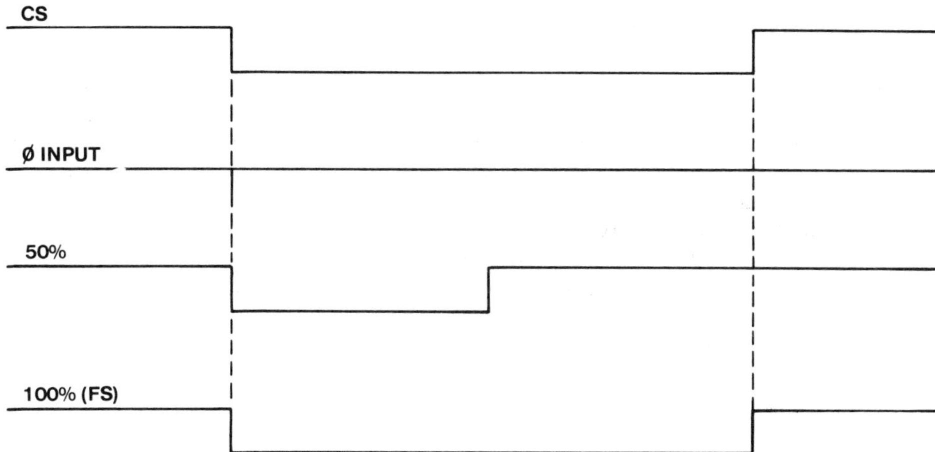



FIGURE 10.14   Pulse-width modulation.

one set of signal lines and monitored by the microprocessor (Figure 10.13). Each V-F converter is assigned an address to distinguish one from another. When the microprocessor wants to read one sensor, the appropriate address is put on the address lines and the tri-state buffer is put into the active state to drive the data line. After the sensor transfers the data to the microprocessor, the device is returned to the high-impedance state by the microprocessor, the next sensor is addressed, and the process continues.

## 10.4.3   Voltage-to-Pulse-Width Converters

Pulse-width (duty-cycle) modulation is another form of serial data conversion. As with voltage-to-frequency converters, the data are transmitted on one wire. The advantage of the pulse-width modulation system is that all the information is transmitted in one pulse, and the ratio of on time to off time of the pulse is measured (Figure 10.14). This technique can be used to control motor speed or the amount of power applied to a system. Pulse-width modulation is used in a manner similar to that used for voltage-to-frequency converter. Instead of measuring the frequency for a fixed period of time, the ratio of the time that the output pulse is in one state to the total pulse width is measured and converted into a digital word.

## 10.5  DATA-ACQUISITION SYSTEMS

A data-acquisition system (DAS) usually denotes a group of electronic
devices that are connected to perform the measurement and quantization of analog electrical signals for digital processing. For the purpose
of this text, the inputs will be assumed to be physical analog inputs
such as temperature, pressure, position, etc.

The trend is to have a microprocessor or minicomputer control
the data-acquisition system and process the data. The basic uses
of data-acquisition systems include four broad areas:

1.  Data logging
2.  Signal analysis
3.  Automatic testing
4.  Process control

### 10.5.1  Typical Systems

Definitions of the key parts of the system include the following:

*Analog Multiplexer*:  This function allows the selection of one of
  several input signals. The multiplexer is digitally controlled
  by the microprocessor or computer.
*Signal Conditioner*:  This function can be performed before or after
  the multiplexer, depending on the type of sensor used. The
  conditioning required might include one or more of the following:
  filtering, linear or logarithmic amplification, peak detection,
  sample and hold, or track and hold.
*Analog-to-Digital Converter*:  The analog-to-digital converter converts the analog input into a digital format.
*I/O Controller*:  The I/O controller generates the system timing
  and controls the memory read/write functions.
*Minicomputer*:  An operational computer system based on a microprocessor chip. It contains memory and is controlled by software.
*Output Buffer*:  The output buffer combines the data with the proper
  signal format for output to specific peripheral devices. The
  output buffer is also called a *peripheral controller*.
*Peripheral*:  Different microcomputer-type output devices, such
  as a line printer, floppy disk, or magnetic-tape storage unit.

### 10.5.2  Data Logging

Data logging involves not only measuring the analog inputs, but
also translating the results into digital signals and storing the data

for further processing or analysis. Figure 10.15 shows the basic system for data logging. These systems can be designed to monitor many channels very accurately and require very little operator attention. The system can be designed to measure many identical channels in a high-volume production system that measures hundreds of units to measure many different parameters and store the data for future analysis.

## 10.5.3  Signal Analysis

Many processes generate time-varying signals from which data must be extracted. The most common method of analyzing information is to translate it into an appropriate Fourier series transform for analysis in the frequency domain. Low-cost computer hardware, coupled with fast Fourier transforms (FFT), make this approach feasible.

### Basic Sampling Concepts

For the data-acquisition system to function, the input signals must be sampled. Sampling may be defined as measuring a continuous function at discrete time intervals. These sampled signals represent some analog parameter converted to a series of discrete values. A typical data-acquisition sampling system is shown in Figure 10.16 and is made up of these major processes:

*Sampling*:   The act of measuring a continuous function at discrete time intervals.

*Quantization*:   Approximating the linear curve by a series of stair-step values called *levels*.

*Digital Processing*:   Evaluating the linear function as a series of discrete values represented by ones and zeros.

*Recovery*:   By the use of a D/A converter, returning the processed information to an analog form.

Besides the required number of bits (or quantization levels), the only other major problems are proper A/D conversion of the digital signal back to analog.

### Types of Signals

The data-acquisition system must be designed to accommodate the type of signal being processed. The nature of the signal must be determined—DC or dynamic, periodic or random, low or high frequency, and noise sources and noise considerations in the signal.

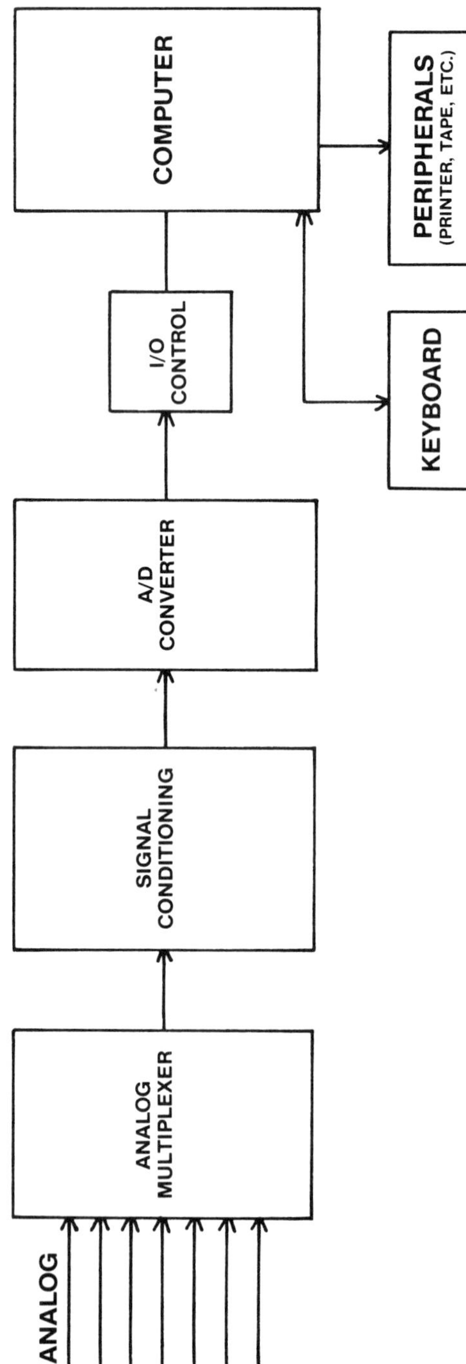


FIGURE 10.15  Data-logging system.

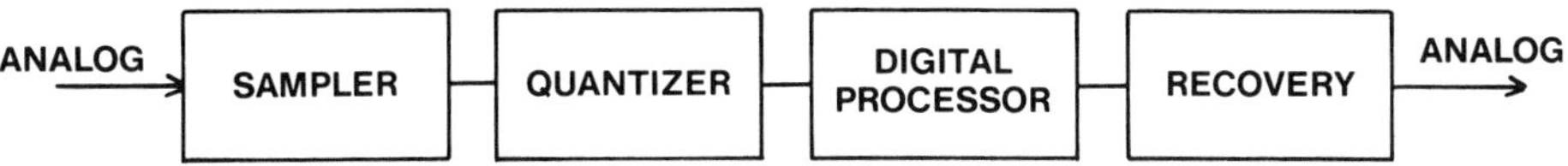



FIGURE 10.16   Data-acquisition system.

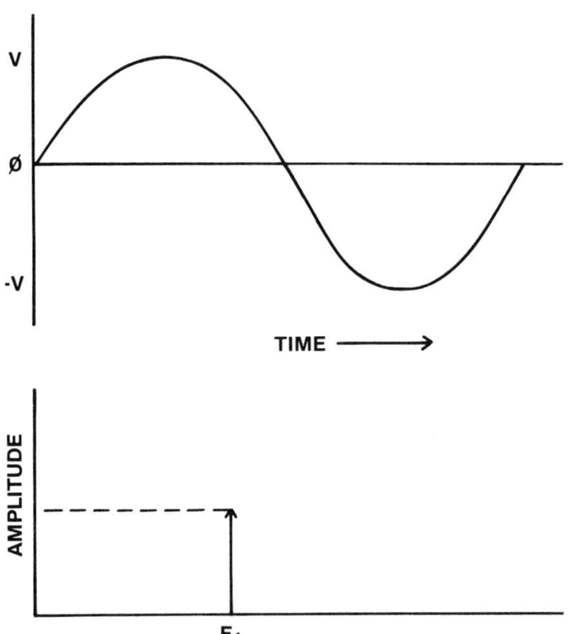



FIGURE 10.17   Time and frequency domain plots of sine wave.

*DC Signals*:   In high resolution systems, measuring a DC signal
can be a major undertaking. The predominant problems are
noise and resolution. Measuring a DC voltage to 0.01 percent
accuracy may sound like a simple task, but impedance matching
and signal loading can be major problems. If a sensor has 5-k$\Omega$
output impedance and the meter used to make the measurement
has 1-M input impedance, the reading will be decreased by
0.05 percent which is out of specifications. This means that
a good part could be rejected due to system loading. The other
major concern when making DC measurements is the amount
of noise generated by the system. The maximum noise must
be maintained less than the smallest signal voltage to be

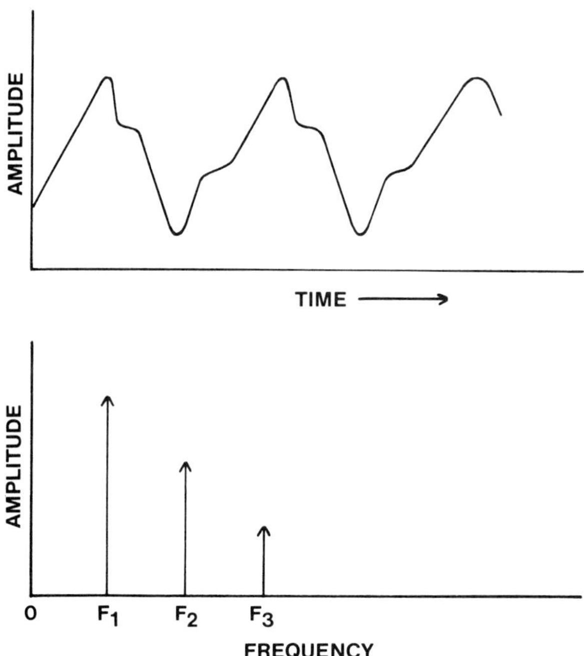

FIGURE 10.18  Typical periodic function.


measured. Filtering is generally used to decrease the noise
level in the system.

*Dynamic Signals*:  The dynamic signal is constantly changing and,
in some cases, in an unpredictable manner. Dynamic signals
may be analyzed in either the time domain or the frequency
domain. The two major classes of dynamic signals are

1.  Deterministic signals—those that have known characteristics
and can be described by mathematics
2.  Random signals—those whose behavior is highly unpredictable

An example of a deterministic signal is a sine wave. The sine
wave has a frequency as well as a time-domain plot (Figure 10.17).
A more complex signal could have a DC component, as well as a
number of discrete frequency components.

As defined, the exact value of a random signal cannot be predicted in advance. Most signals we deal with fall into this class.
The random signal is not repeatable; therefore, its time function is
aperiodic and its frequency plot is a closed spectrum (Figure 10.18).

# 11
# Applications

## 11.1 INTRODUCTION

With improvements in pressure-sensor technology, the price of pressure sensors is decreasing, while the price-performance ratio continues to improve. As the price decreases, the number of applications increases generating a snowball effect: The more sensors manufactured, the lower the manufacturing cost—which decreases the price even further. The number of different markets that have potential applications for low-cost pressure sensors are too numerous to mention.

## 11.2 GENERAL APPLICATIONS

General applications can be divided into three categories: monitoring, control, and physical measurements.

### 11.2.1 Pressure Monitoring

Pressure-monitoring applications make up the majority of the low-cost sensor applications. Simple pressure gages are used to monitor pressure in applications ranging from storage-tank pressures to automotive tire pressures. Many of the mechanical gages are being replaced by low-cost sensors that can be interfaced to electronic monitoring devices.

Today, tire-pressure gages are being dititized using a low-cost
pressure sensor and a low-cost microprocessor. In some higher
priced cars, a new system is being installed that monitors all four
tires as well as the spare tire and gives a warning if the pressure
is outside a predetermined range. These systems are used in conjunction with the new run-flat tire. These new applications all need
pressure sensors that can interface with a microprocessor.

The same trend is happening in process-control systems, where
the low-cost sensors are monitored by a low-cost microprocessor
that processes the data and interfaces to the main computer. This
type of system provides continuous, 24-hour monitoring and can
perform other functions, such as giving trends in usage over a
period of time and establishing an optimum reordering schedule.


## 11.2.2  Pressure Control

Control applications can be divided into two groups: process control,
such as in large chemical plants and refineries, and general control,
such as in HVAC (heating, ventilation, and air-conditioning systems)
and consumer products. Process-control applications are less price
sensitive and generally require more accurate sensors. These applications pay a premium for reliability and minimum system down
time, since each hour a control system is down can cost a company
many thousands of dollars.

General control applications include HVAC systems, where the
heating and air-conditioning ducts are monitored and the amount
of heating or air conditioning is directed to the areas of the building
that require heating or cooling, rather than using one thermostat
to control the entire floor. This type of system improves efficiency
and reduces operating costs by conserving energy. The system
monitors the temperature in each section, as well as the flow in
each duct. The computer can partially block some ducts by controlling
the position of dampers. This closed-loop control system also ensures
an even temperature across the entire floor of the building, rather
than having the side of the building facing the sun too hot and
the other side too cool. These systems are sensitive to sensor price,
due to the number of sensors in a system.


## 11.2.3  Physical Measurements

Pressure can be monitored to give an indication of altitude, depth,
or barometric pressure. Low-cost altimeters and rate-of-climb indicators are being produced for sporting activities, such as hang

gliding and hot-air balloon instrumentation. Portable combination altimeters and barometers are being manufactured for backpackers and mountain climbers.

## 11.3 SPECIFIC APPLICATIONS

In Chapter 8, we discussed how to select a sensor for a particular application, and in Chapter 7, we discussed signal-conditioning the basic sensor. Now we will combine these two chapters and design some general-purpose circuits for specific sensor applications.

In the following examples, some basic circuits using pressure sensors will be discussed. The circuits will cover monitoring, closed-loop control, and physical measurement examples. Monitoring examples will include (1) a sensor and 3-1/2-digit LCD (liquid crystal display), and (2) a sensor and bar-graph display. Examples of closed-loop control applications will include (1) a pressure controller, and (2) a washing-machine controller. The physical-measurement examples will be a barometer/altimeter and a hand-held portable barometer. Each design will discuss sensor selection through amplifier and display module selection.

### 11.3.1 Circuit Design

The following steps must be taken in the design process:

1. Define the end product
   1. Application
   b. Specification
   c. Packaging
   d. Price range
2. Design the circuit, select components
   a. Type of sensor and pressure range
   b. Excitation (reference voltage or current source)
   c. Amplifier stage (if needed)
   d. Output stage (display device)
   e. Hardware
3. Adjustment procedure
   a. Offset
   b. Sensitivity
4. Build and test prototype units

### 11.3.2  Hand-Held Pressure Meter

*Define the Project*

The project will be a hand-held 3-1/2-digit pressure meter, used
to measure dry air, with the following specification:

Range:   0 to 100 psig
Accuracy:   ±2% FS
Temperature:   0°C to 50°C ±3% FS
Power:   9-V battery
Life (battery):   100+ hours
Package:   3.5- × 5- × 1.5-in. molded plastic
Price:   Under $35.00 (parts in lots of 1000)

*Design the Circuit*

A printed circuit board (PCB) will be used to mount the electronics
and display. The pressure sensor can be a low-cost, 100-psi, general-
purpose sensor. The sensor should be PCB-mountable and small
enough to fit in the package dimensions. A small, plastic 0-to-100-
psig sensor with 1/16-in.-ID (inside diameter) pressure ports that
costs $10 each in lots of 1000 will be used. The small-ID pressure
ports will make the pressure connections easier inside the package
by allowing smaller radius bends in the pressure lines and will
allow the use of barbed pressure connections instead of pressure
fittings. A 3-1/2-digit, single-chip ADC (7106) is available from
a number of manufacturers. The 7106 has built-in LCD drivers,
reducing the parts count to one sensor, one A/D chip, one amplifier,
and one LCD. The 7106 has a high-impedance differential input
stage that the sensor can drive directly.

The supply for this circuit will be a battery that will change
slowly with time as the meter is used. The circuit can be designed
to be ratiometric, or a reference voltage can be generated to set
the current source and the reference voltage on the ADC. In this
case, the 7106 is ratiometric to the reference voltage, and a ratiometric current source is easy to design. Since the sensor output
is proportional to the supply current and the A/D output is inversely
proportional to the reference voltage, any change in the battery
voltage will be canceled. Making the circuit ratiometric to the supply
will reduce the parts count, which will save PCB area and cost.

To design the circuit, we need the sensor and 7106 specifications.
The 7106 is a dual-slope, low-power, 3-1/2-digit A/D converter
with a built-in auto-zero circuit and a 200-mV or 2.0-V input range.
This design will use the 200-mV range and drive the ADC directly
from the sensor.

The full-scale reading of the ADC is 1999 or 2000 counts. The
A/D converter sensitivity is ratiometric to the reference voltage,

which allows the full-scale output to be adjusted by changing the reference voltage. When the input voltage is twice the reference voltage, the output will read 1999, and when the input is exactly equal to the reference voltage, the output is 1000. We will set the output of the sensor to be the same voltage as the reference. When 100 psig is applied to the sensor, the ADC will have an output reading of 100.0 psi. The 7106 will take both positive and negative inputs. If +5 psi is applied to the sensor, the meter reads 005.0, and if a pressure 5 psi below atmospheric is applied to the sensor, the reading will be -005.0. The specifications for the 7106 and the sensor are as follows:

*7106 Specifications:*
Zero input reading:   000.0 reading
Linearity error:   ±2 counts
Common-mode rejection ratio:   50 µV/V
Noise:   15 µV
Input leakage (bias current):   10 pA maximum
Scale-factor TC:   5 ppm/°C
Supply current:   0.5 mA typical
Analog common voltage:   2.4 to 3.2 V

*Sensor Specifications:*
Input offset voltage:   ±2 mV
Sensitivity:   100 ± 40 mV
Linearity:   ±0.2% FS
Offset TC:   ±1.0% FS
Span TC:   ±1.0% FS
Stability:   ±0.2% FS year
Excitation:   1.5 mA DC
Bridge impedance:   5000 $\Omega$

The next step is to design the sensor stage, which will include the ratiometric current source. The sensor has an offset specification of ±2 mV which is ±2 percent of full scale. An offset adjustment and a sensitivity adjustment will have to be designed into the circuit. The ADC has no offset adjustment (built-in auto zero), so an offset adjustment will be done at the sensor.

Figure 11.1a is the schematic diagram of the sensor. $R_1$ and $R_2$ are offset temperature-compensation resistors, $R_3$ and $R_4$ are bridge offset-compensation resistors, and $R_5$ is the span-compensation resistor. The values of $R_1$ through $R_5$ are supplied with each sensor. In this design, the specified values for $R_1$, $R_2$, and $R_5$ will be installed on the PCB and $R_3$-$R_4$ will be replaced with a potentiometer (offset adjust). Two resistors will have to be installed for each device—$R_1$ or $R_2$ and $R_5$.

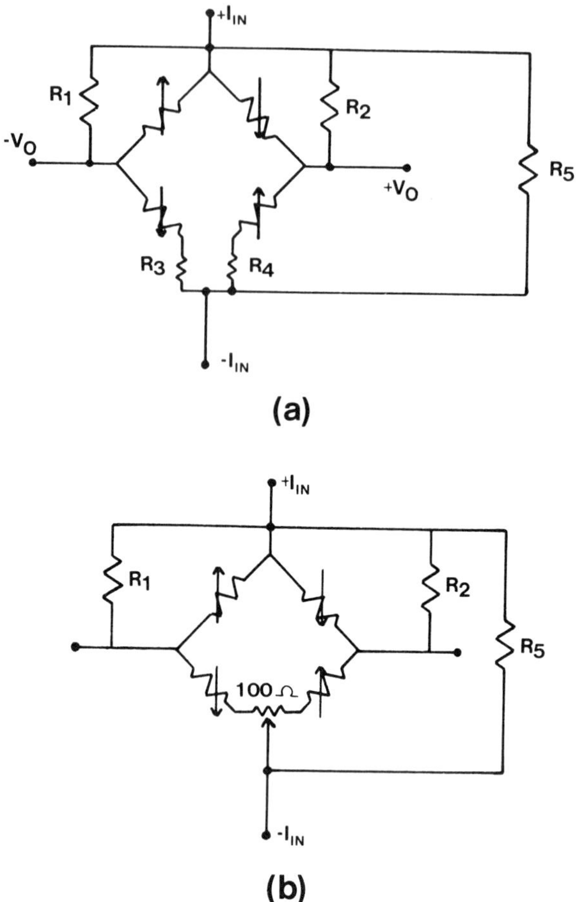



FIGURE 11.1   Sensor schematic.


The ratiometric constant-current source needs to be designed.
The first question is whether there will be enough voltage from
the 9-V battery to operate the sensor at 1.5 mA at the lowest battery
voltage. If the battery is changed when it reaches 7.5 V, the current
source has a compliance of 7.5 V - 1.5 V, or 6 V. At 1.5 mA,
the highest bridge impedance the current source could drive would
be 4 kΩ. The bridge impedance will increase with temperature and,
to ensure that the current source will not saturate, the maximum
temperature impedance has to be used. That is:

$$R_b = 5 \text{ k}\Omega \text{ at } 25°C$$

$$R_b \text{ changes at } 0.21\%/°C$$

$$R_b = (45 \times 2.1 \times 10^{-3})5 \text{ K} + 5 \text{ K} + 5 \text{ K} = 5473 \ \Omega \text{ at } 70°C$$

The maximum current at which the bridge can operate in this circuit is 1 mA at the maximum temperature (70°C). The sensor is specified to have 60 to 140 mV at 1.5 mA and, at 1 mA, the sensor will have 40 to 93.3 mV, for an applied pressure of 100 psi. The full-scale adjustment will be done at the A/D converter. As shown in Figure 11.2, the reference voltage to the current source is generated from a simple voltage divider off the 9-V supply. $R_1$, $R_2$, and $R_3$ are calculated for 1 mA at 7.5-V supply. The current will be a little higher with a fresh battery.

### Calibration

The calibration procedure for this circuit will be to adjust the offset to zero using the offset adjust pot $R_4$, then select a gain resistor $R_G$ to set the reference voltage so that a 100-psi pressure at the sensor reads 100.0 at the meter output (Figure 11.3).

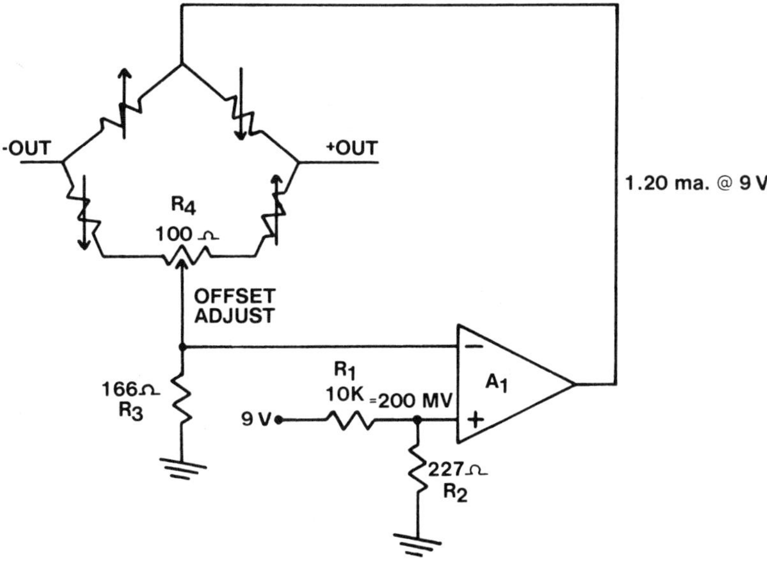


FIGURE 11.2   Sensor and current source.

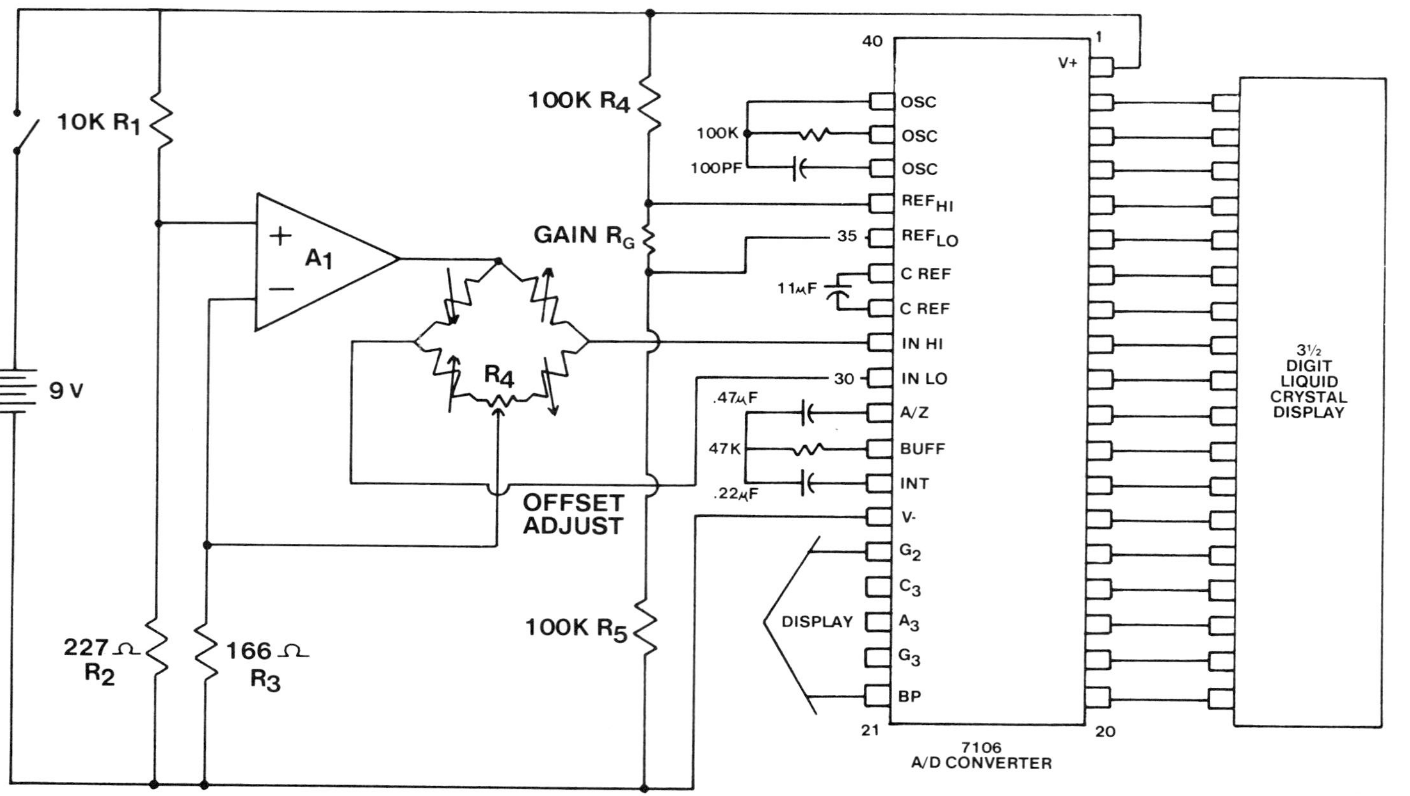


FIGURE 11.3  0- to 100-psi meter.

### 11.3.3  Bar-Graph Meter

In some automotive and appliance applications, a simple bar-graph
meter is used to indicate such parameters as fuel level and temperature. A bar-graph meter is a relatively simple device to design.
A sensor measures the pressure or temperature and generates an
analog voltage proportional to the pressure or temperature. This
analog voltage is used to drive a set of comparators (one for each
bar in the display). The output of the comparator can drive the
display directly or, if an LCD is used, a driver stage would have
to drive the LCD.

*Define the Project*

Design a 10-bar meter to indicate engine oil pressure, with the
following specifications:

Bar-graph meter:  10 bars
Power supply:  12-V DC
Display:  LED (light-emitting diode) bars
Pressure range:  0 50 100 psi
Medium:  Engine oil

*Design the Circuit*

The bar-graph meter design can be divided into three sections:

1.  The sensor and amplifier section
2.  The reference voltage and comparators
3.  The display or LED bars

The sensor (Figure 11.4) will be measuring engine oil that
may contain different additives; to ensure reliability, the sensor
will be an isolated type with a stainless-steel diaphragm. The sensor
output will be amplified to an output voltage of 0 to 4 V for a 0- to
100-psi input pressure, or 4 mV/psi. The sensor specifications are
as follows:

Pressure range:  100 psig
Accuracy (linearity repeatability):  ±0.5% FS
Offset voltage:  ±1 mV
Sensitivity:  100 mV ±2% FS
Temperature:  0.04% FS/°C
Excitation:  10 V ±1%
Input impedance:  5 k$\Omega$
Pressure connection:  1/4-in. NPT male

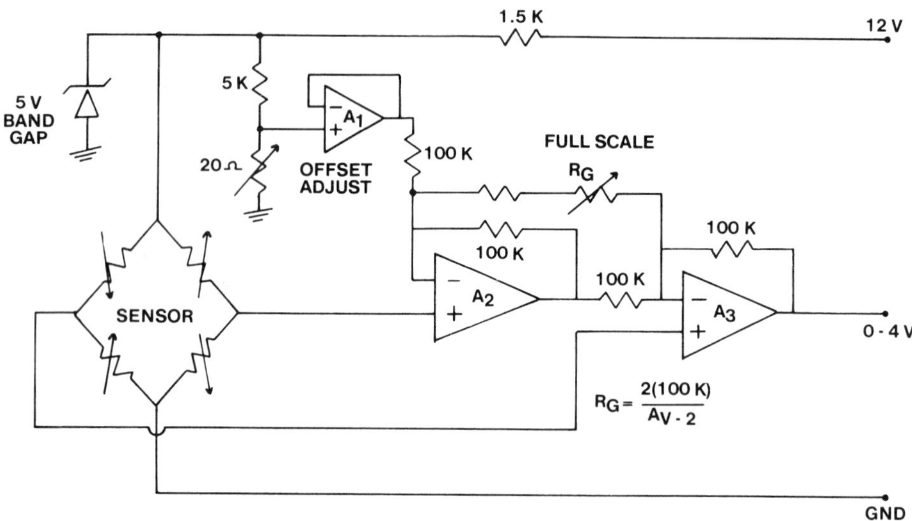


FIGURE 11.4   Pressure sensor.

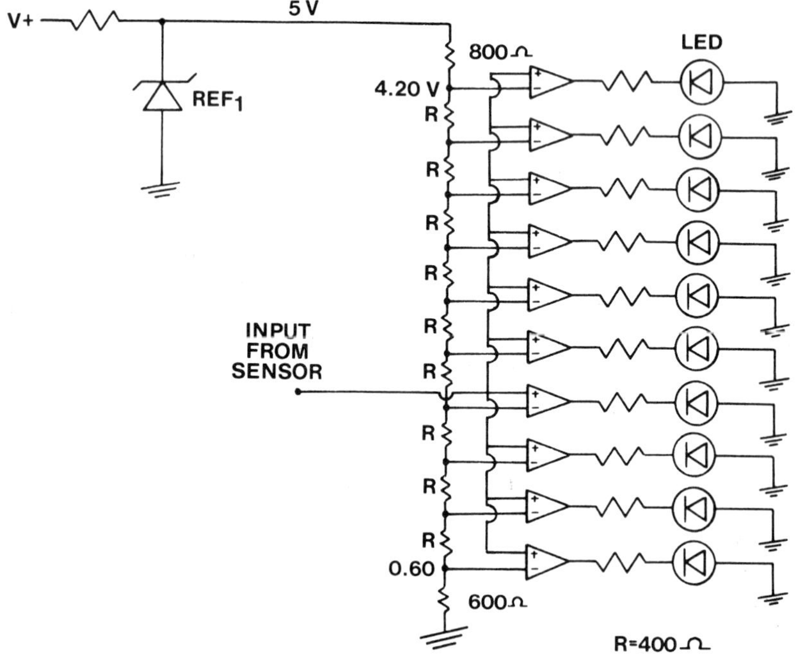


FIGURE 11.5   Reference voltage and comparators.

The pressure sensor has a sensitivity specification of 0.1 mV/V/psi and is operated from a constant voltage source. In this design, the sensor and reference voltage divider will be operated at 5 V. At 100 psi, the sensor will have an output of 50 mV. To get an output of 4 V from the amplifier, the amplifier gain is 80 (400 mV/10 psi). $R_G$ will have to be 2.56 k$\Omega$, or a 2-k$\Omega$ 1 percent resistor and a 1-k$\Omega$ potentiometer in series (full-scale adjustment).

The reference and comparator stage is simply a voltage divider from the 5-V band-gap reference that biases 10 comparators (Figure 11.5). The comparators are 124-type operational amplifiers; they are quad amplifiers, so the entire circuit takes only four amplifier packages.

As pressure is applied to the sensor, the amplifier output increases until the first comparator trip point is reached (at 10 psi). The first comparator changes state, turning on the first bar. This process continues, with another bar being turned on every 10 psi, until the last bar is turned on at 100 psi. The sensor has a ±2-mV offset-voltage specification, which will have to be nulled out to ensure that the first comparator will change state at 10 psi. The sensor offset (±2 mV) times the amplifier gain gives an error at the amplifier output of $\pm 2 \times 80 = \pm 160$ mV. To compensate for the offset, the reference divider will have to be offset by 200 mV. The first bar will turn on at 600 mV and the tenth bar at 4.20 V.

### Calibration

The offset is adjusted first; with 10 psi applied to the sensor, the offset adjust pot is set so that the first segment just turns on. The sensitivity is adjusted by applying 100 psi to the sensor and adjusting the full-scale adjust pot so that the tenth segment just turns on. As the pressure is increased from 0 to 100 psi, one segment should turn on every 10 psi.

## 11.3.4  Pressure Controller

A circuit is required that will maintain the pressure in a pressurized instrument housing at 20 psig above atmospheric pressure.

### Define the Project

The pressure controller will have the following specifications:

Control range:   20 ±1.5 psig  
Temperature range:   0°C to 70°C  
Thermal effects:   ±3% FS

Medium:  Dry nitrogen
Power supply:  12 V
Outputs:  High, low valve drivers

*Design the Circuit*

The circuit will control the pressure in the housing by turning
on a vent valve when the pressure gets too high. When the pressure
gets too low, the circuit will turn on a valve that purges the housing
from a tank of dry nitrogen. As long as the pressure is between
18.5 and 21.5 psig, both valves remain closed. The circuit is designed in two stages: the sensor and amplifier stage and the comparator and valve-driver stage.

Since the sensor will be measuring clean, dry nitrogen, a standard
pressure sensor will work. The first question is which pressure
range should be used—the standard ranges are 15 psi and 30 psi.
Either range could be used; the 15-psi sensor has twice the sensitivity and will perform better over temperature, while the 30-psi
sensor has a higher overpressure specification. In this application,
sensor linearity is not critical, since the sensor is operating at
one pressure point. The 15-psi sensor has more advantages.

The sensor specifications are as follows:

Offset voltage:  ±2 mV
Full-scale output:  100 ±40 mV
Thermal accuracy—offset:  ±1% FS
Thermal accuracy—span:  ±1% FS
Repeatability:  ±0.05% FS
Long-term stability:  ±0.3% FS
Excitation:  1.5 mA
Overpressure:  3 × full-scale pressure

The circuit will be operating at 20 psi; the sensor has a typical
specification of 6.67 mV per psi or 133.3 ±53 mV at 20 psi.

Figure 11.6 shows the sensor, current source, and amplifier
stage. $R_G$ is calculated for each sensor to give a sensitivity at the
amplifier output of 250 mV/psi. With this sensitivity, the amplifier
output will be 5 V at 20 psi. The comparators and valve drivers
can be designed, now that the sensor stage has been defined. One
comparator will be needed for the low set point and one for the
high set point. When driving valves, some hysteresis should be
designed into the comparator to keep the noise in the circuit from
making the valves chatter.

Figure 11.7 shows the reference amplifier, comparators, and
valve drivers. Amplifier $A_4$ amplifies the 1.235-V reference voltage
to 6 V as a reference for the comparators. Operational amplifiers

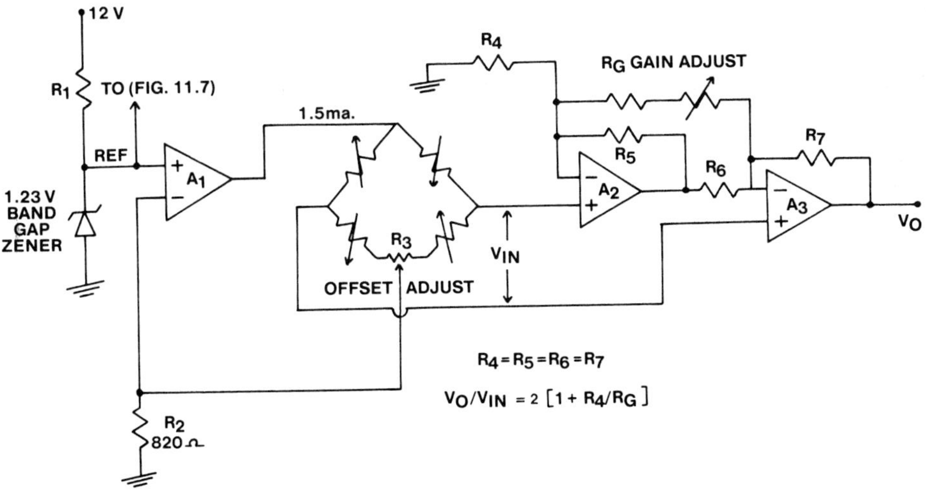


FIGURE 11.6  Sensor and amplifier.

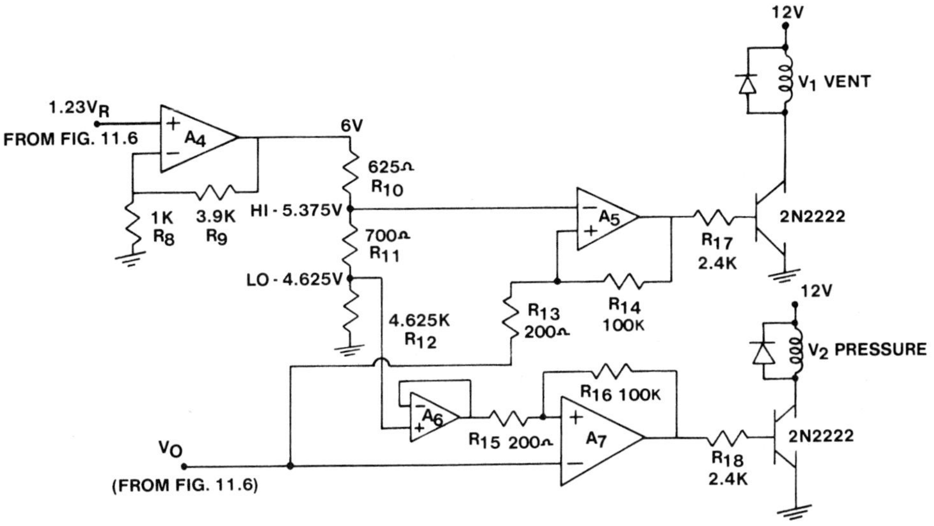


FIGURE 11.7  Pressure controller.

used as comparators $A_5$ and $A_6$ are 124-type operational amplifiers
with ±10 mV of hysteresis to reduce noise chatter. The ratio of
feedback resistor to input resistor is 500 to 1; with a 100K feedback
resistor, the input resistor is 200 $\Omega$. Amplifier $A_6$ buffers the voltage
divider to provide a low-impedance voltage source to drive $R_{15}$.
Each comparator drives an NPN transistor that provides the power
necessary to operate the control valves.

*Calibration*

This circuit cannot be calibrated directly, due to the hysteresis
around the comparators. The circuit can be calibrated one of two
ways: (1) $R_{14}$ and $R_{16}$ could be left out until after the calibration,
reducing the hysteresis to 0, or (2) the sensor amplifier output
could be set to 4.625 V at 18.5 psi with the offset adjust and,
if the gain has already been set to 250 mV/psi, the other switch
point will be at 21.5 psi.

## 11.3.5  Washing-Machine Controller

A microprocessor-controlled washing machine will use a number
of sensors to control the amount of water and the water temperature
and to measure and dispense the right amounts of soap and bleach
during the cycle. In the future, the washing machine could be
designed to control the entire washing cycle. The clothes to be
washed would be placed in a special basket that is just large enough
to hold the maximum load for that particular washer. The clothes
and basket would be placed on the washer lid and weighed by a
load cell built into the lid; this number would be fed to the microprocessor, which would calculate the right amount of soap, conditioner, and bleach to be used, as well as the right amount of water
and temperature for this load. Two pressure sensors would be used
in a system like this: one in the lid to make a load cell and weigh
the clothes and a second to control the amount of water used in
the wash and rinse cycles. Having the microprocessor in the system
makes the load-cell design much simpler, because the microprocessor
can auto-reference the pressure sensor by taking a measurement
with no basket in place and storing that number as the zero reference
number for the following readings.

*Load-Cell Design*

The load cell could be a series of pistons that the basket is placed
on. Figure 11.8 shows one possible way that the lid could be designed.
The pistons could be special rolling diaphragms with a hard rubber

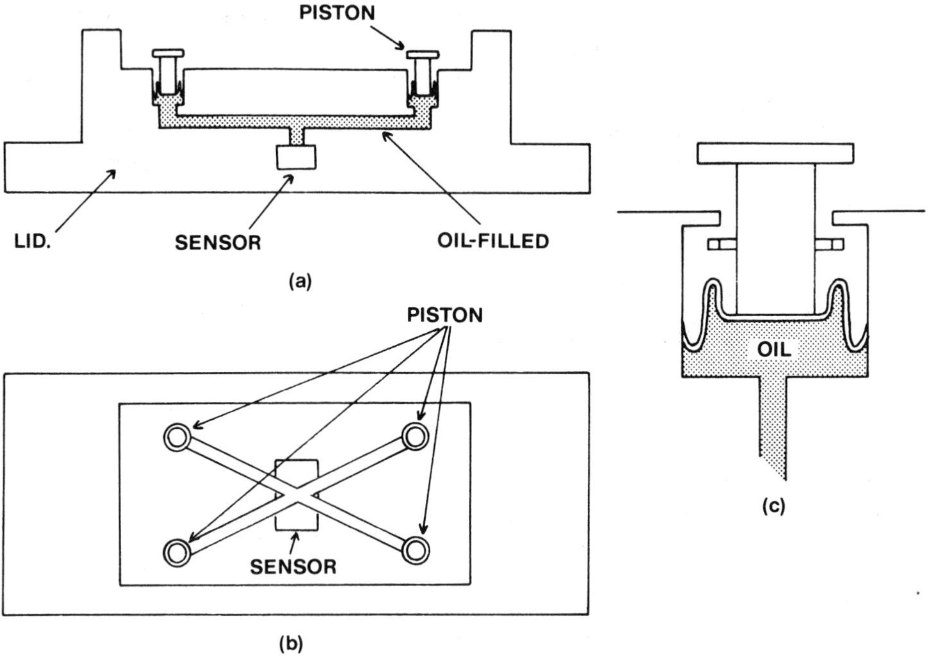



FIGURE 11.8   Load cell.

cylinder molded in their centers. The rolling diaphragm eliminates
the problem of friction associated with a standard O-ring type of
piston and still allows a large stroke. Figures 11.8a and 11.8b show
the pistons and sensor in the lid, and Figure 11.8c shows the rolling
diaphragm. The critical factors in this design are that the sensor
is not allowed to saturate and the load cells do not bottom out.
The load cells can be designed with a large enough stroke to ensure
that any expansion of the oil used to couple the sensor to the piston
will not overpressure the sensor. Stops would be designed into
the piston housing to limit the travel. The weight of the clothes
is measured by reading the output of the sensor and multiplying
by a constant to convert from psi to pounds. From Chapter 2,

$$F_2 = F_1 \times A_1/A_2$$

where $F_1$ is the applied force

$F_2$ is the force or pressure applied to the sensor

$A_1/A_2$ is the ratio of the two areas (1 in this case)

The area of the piston will determine the pressure developed at the sensor. From Chapter 2, when 10 psi is applied to a piston with an area of 1/2 square inch, a pressure of 5 psi will be developed at the sensor.

When designing consumer products, price will be a prime consideration. Components used in consumer products are generally low cost, but are used in high volume. The factors that have to be considered when selecting the sensor are

1.  Media compatibility
2.  Package
3.  High-level/Low-level output
4.  Overpressure
5.  Price

*Media Compatibility*   The sensor will be isolated by an oil column that allows the use of a low-cost, unisolated sensor. The only compatibility question involves the oil and the sensor materials. If the sensor has a coating of silicon gel over the sensing chip, an oil that is compatible with the gel and package material must be selected.

*Package*   Since this is a high-volume application, there will be a number of options available:

1.  Standard package
2.  Modified standard package
3.  Custom package

In most applications, using a standard transducer package will be the least expensive; the problem is that if the lid is not designed around the transducer package, it may be difficult to make pressure connections and/or find a transducer the right size. Some companies will modify the standard package to fit the application if a large order is involved. The most common solution is a custom package; designing the package to be easy to fill with oil and be a functional part of the lid would be most cost effective. The package drawings and sensor specifications would be sent to a number of sensor manufacturers to establish more than one source for the sensor. The part could be molded out of plastic and have all the passages for pressure connections built in.

*High-Level/Low-Level Output*   If a transducer with the right output voltage range can be found, buying a high-level output would be less expensive and have less noise problems than trying to use a low-level signal. All the sensor outputs will probably be multiplexed and converted by an A/D converter that will be controlled by the microprocessor.

*Overpressure*   The sensor should have an overpressure specification of at least four or five times full scale, because people will drop the basket in the lid.

*Price*   Throughout this discussion, price has been mentioned. The price of the sensor includes

1.  The sensor itself
2.  Pressure connections
3.  Mounting hardware
4.  Pressure tubing
5.  Labor (assembly, testing, calibration)

If a high-priced pressure connector has to be used with a particular transducer, or special mounting hardware needs to be machined, these things need to be added to the price of the pressure transducer.

The transducer specifications for the load cell are as follows:

Pressure range:   0 to 15 psig
Output voltage:   0 to 5 V
Linearity:   ±0.2% FS
Temperature error (zero):   ±5% FS
Temperature error (span) (temperature 0°C to 50°C):   ±2% FS
Power supply:   12-V DC

The package will be a custom molded-plastic package, with pressure connections built in, that snaps into the bottom of the washer lid.

## Level Sensor

The second sensor used in the washing machine is a water-level sensor that measures the amount of water used for each cycle of the washing process. This sensor can have relaxed specifications as well, due to the use of the microprocessor to auto-reference and calibrate the sensor. This sensor will also use a rolling diaphragm and oil fill to isolate the sensor from the wash water. A screen will be used over the diaphragm to protect it from sharp objects. The transducer will have the same specifications as the load-cell transducer, with the exception that the pressure range is 1 psig (27 inches of water) full scale. The sensor is mounted on the bottom of the tub and will be subject to all the vibration and movement of the tub. The water-level sensor specifications are as follows:

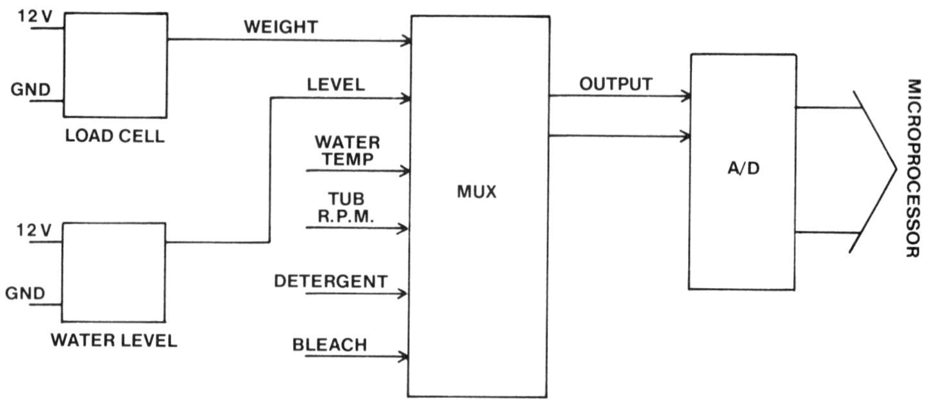


FIGURE 11.9   Washing-machine system.

Pressure range:   0 to 1 psig
Output voltage:   0 to 5 V
Linearity:   ±0.2% FS
Temperature error (zero):   ±5% FS
Temperature error (span) (temperature 0°C to 70°C):   ±2% FS
Power supply:   12-V DC

Figure 11.9 shows the two transducers and the multiplexer used
to interface to the microprocessor.


### 11.3.6   Barometer

Many computer-controlled systems are affected by barometric pressure. A barometer that will interface directly to the computer allows
automatic calibration to correct for barometric pressure changes.
A number of units could be used to track weather highs and lows
as they pass over the area. It is being found that changes in barometric pressure can also have an effect on some critical medical
tests. A barometer built into the equipment can make any necessary
calibrations to the equipment automatically as the barometric pressure
changes. We will discuss two barometer designs: The first will be
a barometer/altimeter that interfaces to a computer, and the second
will be a hand-held (portable) barometer with an LCD.


*Definition*

Designing a barometer has some unique problems. A small absolute
pressure change is measured at the local barometric pressure, which

could be from 8 psi (16,000 ft above sea level) to 15 psi (560 ft
below sea level). Barometer readings are normalized to sea level,
so a barometer in Denver will read the same as a barometer in San
Francisco. In a system that is sensitive to absolute barometric pressure, the computer can adjust for altitude (large change), as well
as for changes in barometric pressure, with one reading. When
the system is installed, the first reading will compensate for altitude
changes, while later readings will make fine adjustments for barometric changes.

*First Design*

The first design will be an absolute-pressure device with enough
resolution to compensate for barometric pressure changes. Two
popular units for barometric pressure are inches of mercury and
millibars. One standard atmosphere is 29.92 in.Hg or 1013.8 millibars
(mbar). The barometer should have enough resolution to resolve
0.01 in.Hg or 1 millibar (12-bit).

*Component Selection*   Component selection is critical in a barometer
design. The sensor, voltage or current source, amplifiers, and
resistors should be selected for stability and low noise to ensure
that the barometer will have enough resolution to measure both
altitude and changes in barometric pressure. The altitude will have
a pressure range from 8 to 15 psi, while the barometer will have
a range of ±0.5 psi from the altitude reading.

*Design*   The device will operate from 8 to 15 psi (16.29 to 30.54
in.Hg or 551.6 to 1034.2 mbar). If a 15-psia sensor is used, it
will operate from 1/2 scale to full scale. The device will operate
close to full scale in most applications; the errors will be the sum
of the offset and span errors. Using half the specified span will
improve linearity error approximately two to one.

*Sensor Excitation*   The design of the current or voltage source
will be just as critical as the sensor span errors. The critical components are

1.   Reference device
2.   Voltage/current-setting resistors
3.   Operational amplifier

    Any change in the reference voltage will have a one-to-one
effect on the full-scale output. Parameters such as temperature
coefficients, stability, and noise should be considered when selecting
the reference device.
    The same considerations apply to the current-setting resistors
in the current source and the gain-setting resistors in the voltage

source. The voltage source is a little less critical for absolute
resistor values, but the resistor ratios are important. If all resistors
have the same sign and percentage error, they will cancel. Considerations for the operational amplifier include temperature coefficients,
bias current, long-term stability, and noise.

The individual errors will be small, but the sum of all the errors
can be significant—especially long-term errors like stability that
cannot be compensated.

*Sensor*   The sensor is the critical component in the system; it will
determine the ultimate accuracy of the system. There are many
15-psi absolute-pressure sensors available, and the package and
specifications will vary directly with the selling price. The mass-
produced plastic package with general-purpose specifications will
be the least expensive, while the metal-can (tight specification,
more testing) sensors have a higher piece price. For this application,
the following sensor will be used:

Pressure range:   15 psia
Sensitivity:   100 ±25 mV
Zero offset:   ±2 mV
Linearity and hysteresis:   ±0.1% FS
Temperature effects (offset 0°C-70°C):   ±0.5% FS
Temperature effects:   ±0.5% FS
Bridge impedance:   5 k$\Omega$
Long-term stability:   ±0.1% FS
Package:   TO-8
Excitation:   1.5 mA

The specifications for the band-gap reference are

Reference voltage:   1.235 V
Voltage change with current (1 to 20 mA):   10 mV
Dynamic impedance:   0.6 $\Omega$
Wide-band noise:   60 $\mu$v/peak-to-peak
Long-term stability:   20 ppm/1000 HR

The circuit for the barometer/altimeter will consist of a standard
operational-amplifier current source, sensor with compensation resistors, and amplifier stage to set the slope (Figure 11.10).

*Current Source*   Amplifier $A_1$ and the band-gap reference make up
the current source, and resistor $R_1$ is a low-TC precision resistor
that sets the bridge current. $R_1$ sets the current through the band-
gap reference. If the supply voltage is allowed to change significantly, resistor $R_1$ can be replaced by a constant current source.

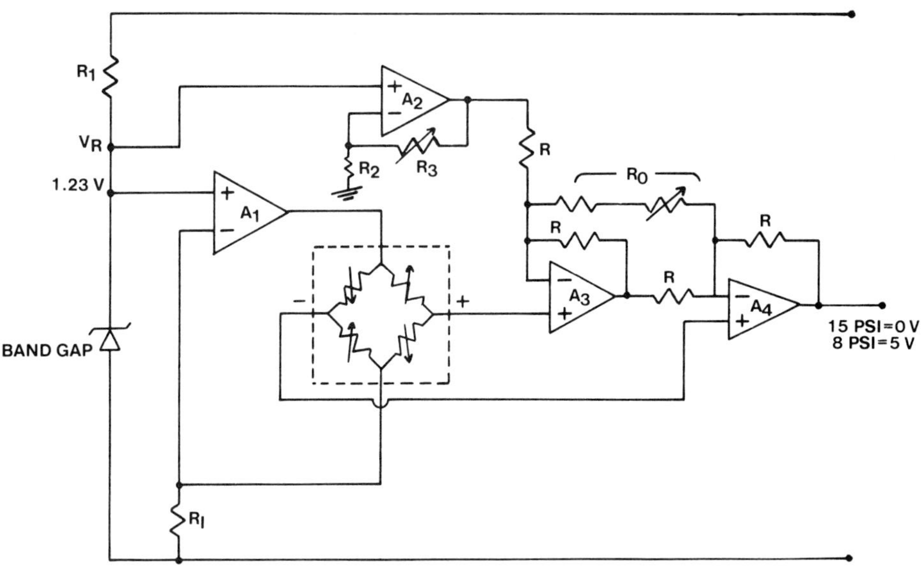

**FIGURE 11.10**  Barometer amplifier.

*Amplifier Stage*  Amplifiers $A_3$ and $A_4$ make up the standard instrumentation amplifier. In this application, the sensor output is inverted, since 15 psi corresponds to sea level and 8 psi is 15,000 feet. The amplifier output will be 0 V at 15 psi and 5 V at 8 psi. Amplifier $A_2$ level-shifts the output by biasing the instrumentation so that full scale from the sensor is 0 V at the output. Amplifiers $A_1$ through $A_4$ are low-noise precision amplifiers.

*Analog-to-Digital Converter*  To get the necessary resolution, a dual-slope A/D converter with an input range of 0 to 5 V will be used to convert the analog output of the sensor to a digital format with which the microprocessor can interface. In this application, there is no display; the computer reads the barometric pressure directly and generates a correction signal for the process. If there were a display, a lookup table would be used to convert barometric pressure to a corresponding altitude reading. The sensor and ADC are calibrated to 0 V at 15 psia by adjusting $R_3$. The sensitivity is adjusted by setting $R_0$.

## Second Design

The hand-held device has the same basic design for the current source, sensor, and amplifier selection. Since the device will be

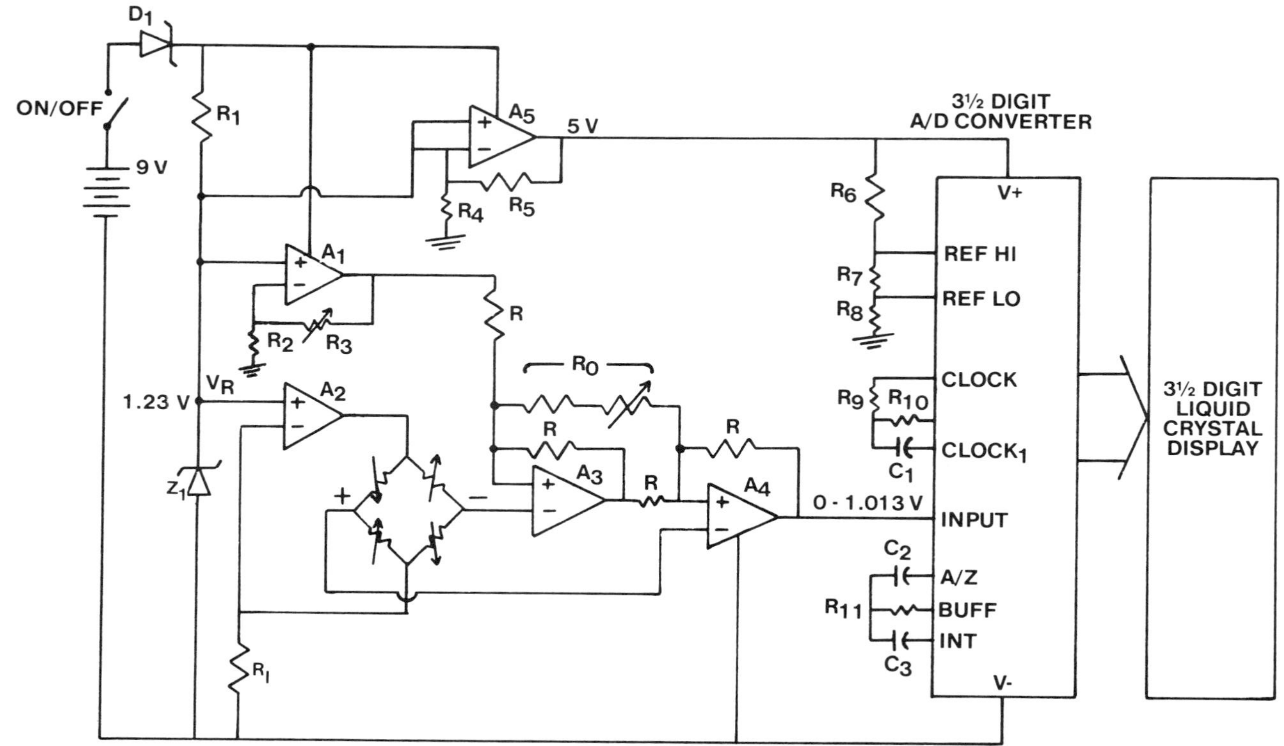



FIGURE 11.11   Hand-held altimeter/barometer.

battery-operated, power consumption must be considered to ensure
a reasonable battery life. A standard alkaline 9-V battery will supply
about 600 mA-hours. If the total current drain is between 3 and
4 mA, the battery will last between 150 and 200 hours (6-1/4 to
8-1/3 days of 24-hour use). The device will have a dual-function
altimeter/barometer. A calibration reading is taken and recorded
at a known altitude. The device would be calibrated in millibars,
and a lookup table supplied with the meter could be used to convert
from pressure to altitude. At a constant altitude, the barometric
pressure would be noted and, as time passed any changes in barometric pressure would indicate a front passing through the area.

*Amplifier*  The same amplifier design will be used (Figure 11.10).
In this application, the amplifier will be selected for low power
consumption as well as low noise and high stability.

*Analog-to-Digital Converter*  The A/D converter will be decimal
rather than binary. A binary-coded 12-bit A/D converter has a
resolution of 1 out of 4096; a decimal-coded 3-1/2-digit A/D converter
has a resolution of 1 out of 1999. For a battery-operated device,
3-1/2-digit CMOS A/D converters with built-in display drivers are
available in one package. The CMOS construction allows low power
consumption with moderate speed. The converter and display will
draw about 1 mA from a 9-V battery. The specifications for the
A/D converter are as follows:

Conversion:  Modified dual slope
Resolution:  3-1/2 digits
Zero input reading:  0000
Linearity:  ±1 count
Common-mode rejection:  5 μV/v
Noise voltage (peak-to-peak):  10 μV
Input bias current:  10 pA
Zero-reading temperature coefficient: 1 μV/°C
Scale-factor temperature coefficient:  5 ppm/°C
Supply current: 120 μA
Operating temperature range:  0°C to 70°C
Reading rate:  2.5 readings/s

In this application, the display will be calibrated in millibars
(1 standard atmosphere = 1013 millibars). The full-scale input of
the analog-to-digital converter is 2.00 V for 1999 counts. This
application will use half the A/D range, or 1.013 V at 1 atmosphere
and a sensitivity of 1 mV per millibar. At sea level, a change of
1 millibar is equal to a change of 27 feet, which would be enough

resolution for a backpacker. A lookup table can be used to correct for the nonlinearity of pressure to altitude up to 15,000 feet.

Figure 11.11 shows the system. Amplifier $A_2$ sets the sensor current, $A_1$ is offset adjust ($R_3$), $A_3$ and $A_4$ are instrumentation amplifiers, and $A_5$ generates a stable 5-V supply from the band-gap reference. As the battery voltage decreases during use, amplifier $A_5$ maintains a constant 5 V until the battery reaches 7V. At 7 V, amplifier $A_5$ saturates and the 5 V decreases—time to replace the battery.

The circuit has two adjustments. $R_0$ is adjusted first for a sensitivity of 1 mV/millibar. Once the sensitivity is set, potentiometer $R_0$ is sealed. The second adjustment is to set the offset to read the absolute barometric pressure. This potentiometer could be made accessible to the user to adjust the reference altitude reading. The climber would know the starting altitude and could adjust the meter to read the same as the lookup table reading for that altitude.


## 11.4  CONCLUSION

By combining the fundamental circuits, constant current source, constant voltage source, and instrumentation amplifier gain stage, any number of circuits can be designed. The user's imagination is the only limit to the possibilities.

# 12

# Future Trends



## 12.1  INTRODUCTION

Pressure and temperature are fundamental physical measurements required by many industries. Medicine, automobiles, process control, and sporting equipment are a few applications that measure pressure and temperature, and each application has special accuracy and packaging requirements.

In the past, low-cost sensors were mechanical (bellows, Bourdon tubes) and used mechanical displays. These devices are used as indicators, but cannot be interfaced to other electronic control equipment. Other techniques, such as potentiometric and LVDT (linear variable differential transformer), have electrical outputs that can be interfaced to other electronics, but are limited in performance and/or package size. The high-accuracy applications such as process control use capacitive transmitters. Most process-control transmitters are expensive, highly accurate, and designed for high reliability.

The age of the computer is generating a need for new, low-cost sensors that can be interfaced to a computer, allowing the system to measure and control physical parameters. As the price of the microprocessor and sensors continues to decrease, the number of potential applications increases rapidly.

Semiconductor batch processing is producing accurate, low-cost silicon pressure sensors that can be used to provide real-time data to the microprocessor. As a result, microprocessors are being designed into cars, household appliances, audio equipment, and watches.

   As the number of practical applications increases, the cost
of sensors is going down and applications are being improved. More
of the temperature-compensation and calibration tasks that were
left up to the user in the past are being incorporated onto the
sensor chip. This has a dramatic effect on the amount of labor
(cost) the end user must put into installing and calibrating sensors.
To summarize, the following topics will be discussed in this chapter:

1.   Evolution of the silicon pressure sensor
2.   On-chip compensation
3.   Future developments


## 12.2   EVOLUTION OF SILICON SENSORS

### 12.2.1   The First Silicon Sensors

The fundamental piezoresistivity of silicon was discovered in the
early 1950s by the same research teams that were working on semiconductor technology. At the same time as they discovered the
diode and transistor, they also noted that silicon was sensitive
to external strains.

   The first commercial applications for silicon strain gages were
in the late 1950s and early 1960s. Silicon resistors were used to
measure the strain induced in a metal diaphragm when pressure
was applied to one side of the diaphragm. Individual silicon resistors
attached to a metal diaphragm sense the strain and generate an

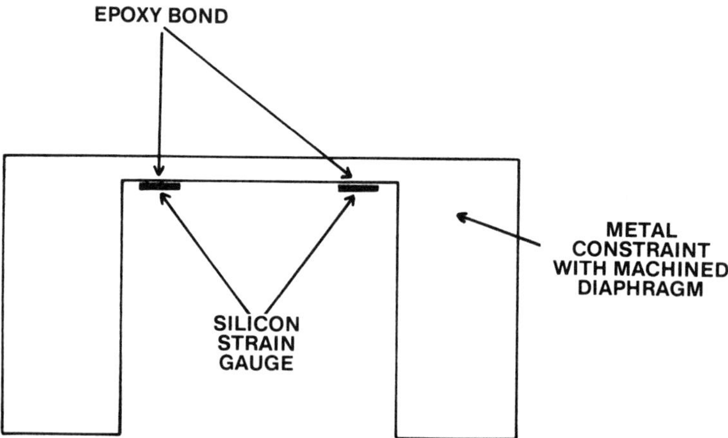


FIGURE 12.1   Bonded strain gage.

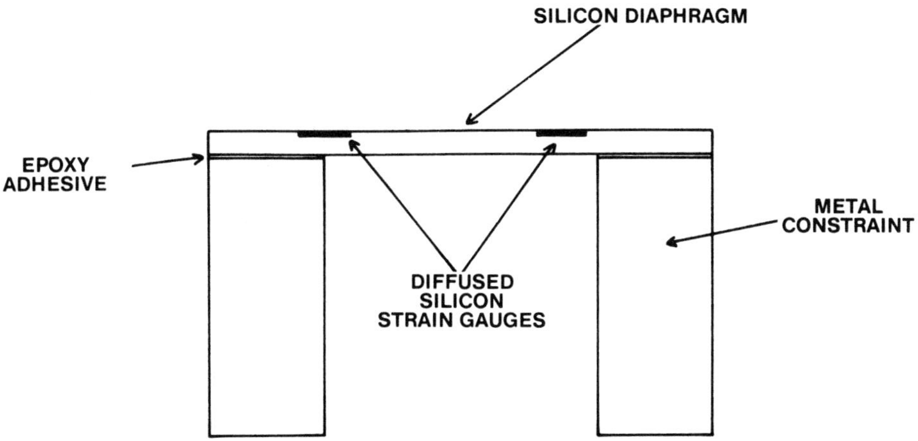



FIGURE 12.2  Diffused-silicon diaphragm.

electrical output by unbalancing the bridge. Similar sensors are manufactured today (see Figure 12.1).

### 12.2.2  Second-Generation Sensors

Second-generation sensors improved performance by making the diaphragm of silicon and diffusing the resistors into the diaphragm. This step removed the adhesive between the sensor and diaphragm, which had caused many problems, including instability and high temperature coefficients (Figure 12.2).

### 12.2.3  Third-Generation Sensors

Third-generation sensors used even more silicon. The cavity was machined into the silicon, creating the diaphragm and using the silicon wafer to support the diaphragm. This step eliminated the thermal mismatch of the silicon to the metal housing. As shown in Figure 12.3, the cavity creating the sensing diaphragm and support are sealed to a second wafer, making a reference cavity for absolute sensors and making the structure stiffer by supporting the silicon cavity from the back side. These are the first parts that used micromachining to batch-process the sensors in wafer form, thus creating the potential to significantly reduce the production costs at high volumes. By processing 300 to 400 sensors per wafer and 10 to 20 wafers per run, thousands of sensors can be

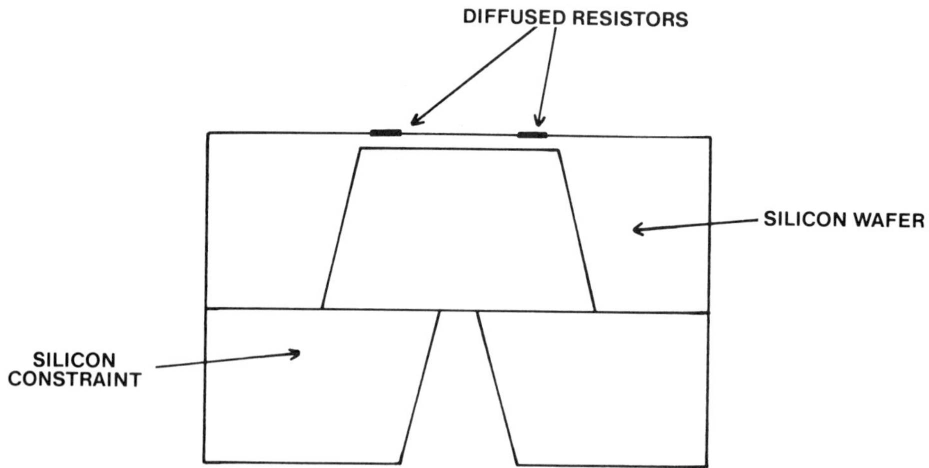


FIGURE 12.3   Etched-cavity sensor.

processed at one time. Batch-processing does not have a large
price advantage if only a few sensors are produced. During this
generation, applications such as automotive (manifold vacuum sensors
and barometric pressure sensors) and medical (disposable blood-
pressure sensors) became feasible. For years, the process suffered
from the chicken-or-the-egg syndrome; users wanted low price,
but could not commit to the high volume, while sensor manufacturers
could make low-cost sensors if the volume were high enough. The
automotive applications looked like a high-volume use, but disposable
blood-pressure sensors turned out to be the first high-volume sensor
application.

## 12.2.4   Fourth-Generation Sensors

The fourth generation has been devoted to performance advancements
and price reductions by improving the basic processing techniques
used to produce the sensors. As a result, the sensors now cost
less and have significantly better specifications than did early
devices. In the early devices, the wafers were sealed together
by using a glass frit and heating the device until the glass reflows,
sealing the two wafers together. The glass interface caused some
stability problems, as well as temperature problems. During the
fourth phase, it was found that silicon can be electrostatically bonded
to Pyrex®, eliminating the glass interface and making a molecular
bond between the silicon and Pyrex®. This process improved the
temperature and stability problems of the early devices.

Improvements have also been made in the cavity-etch process, resulting in more precise diaphragm thickness and more uniform cavities from device to device as well as from wafer to wafer. During the fourth phase, the resistors were ion-implanted into the silicon and, to make small, low-pressure sensors, bossed diaphragms were introduced to concentrate the diaphragm strain in the resistor locations. Automated packaging improvements were introduced, using multiple lead frames similar to IC packaging techniques. A strip containing 10 or 12 units is automatically die-attached and bonded on automated die bonders. This reduces the labor required for each device and results in more uniformity from device to device. The price of sensors has reached a level where some medical applications, such as blood-pressure monitoring, are using disposable sensors that are used one time and discarded along with the plastic tubing. Sterilizing the sensor is more expensive than replacing it.

## 12.3 ON-CHIP COMPENSATION

Pressure sensors are available in one of three classes:

1. Silicon chip or packaged chip with no testing other than a functional test
2. Sensor that has been tested and for which the compensation values are supplied with the sensor
3. Sensor that has been tested and to which a compensation ceramic has been added to provide the compensation and sometimes calibration resistors

The next step in the evolution of the pressure sensor was to bring more functions onto the chip, such as temperature compensation and calibration. To accomplish this, some new processes had to be developed:

Developing and combining processes for piezoresistors and low-temperature-coefficient/laser-trimmable resistors
Designing capabilities to trim chips in either wafer or packaged form

One of the problems encountered in designing compensation networks is that the resistors have to cover a large range of values. A 5-k$\Omega$ bridge would have resistor ranges as follows:

Offset resistor:  0 to 100 $\Omega$
Span-compensation resistor:  10 to 25 k$\Omega$
Offset temperature compensation:  200 k$\Omega$ to 5 M$\Omega$

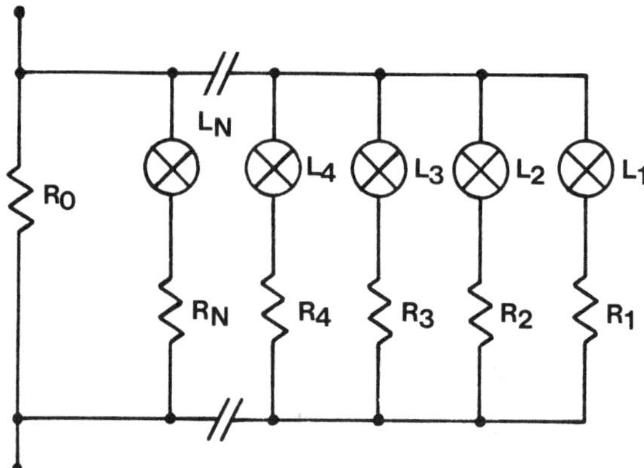

FIGURE 12.4  Digital-trimming network.

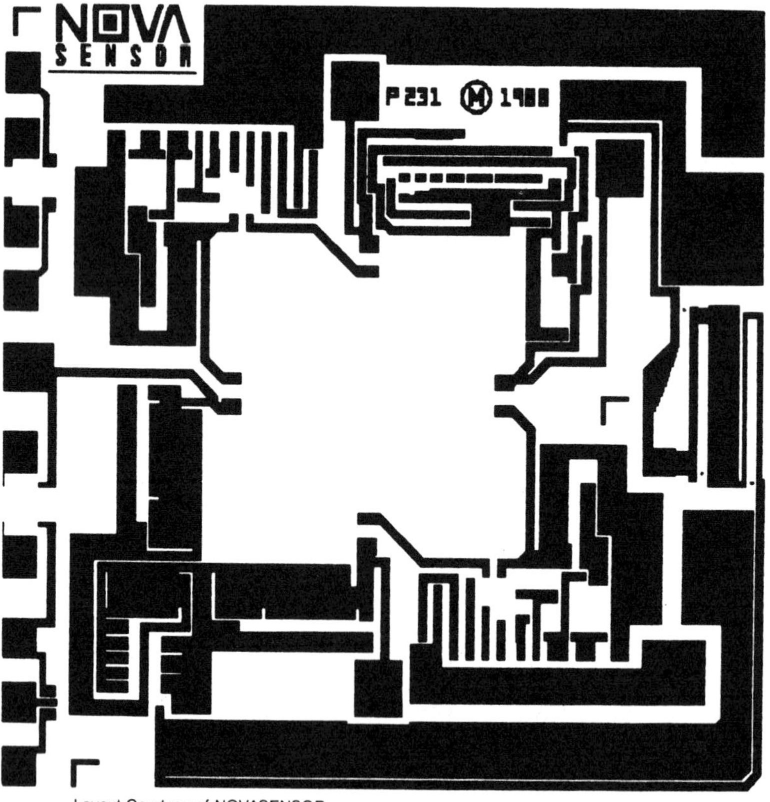


FIGURE 12.5  Compensated and calibrated sensor chip.

A new thin-film-resistor process with low temperature coefficient
was developed to produce resistors with less than 100-ppm temperature coefficients (equivalent to a metal-film discrete resistor).

To achieve the large range, resistor networks were designed
that can be laser-trimmed to the desired value. The technique used
to trim the resistors is known as *link blowing*. Metallized links
are evaporated by laser energy, making the link act like a switch
in a binary ladder network. The impedance of the network can be
set by connecting and disconnecting parallel resistors. The resolution
of the trim is directly proportional to the number of links. Figure
12.4 shows a typical resistor network that could be used to trim
sensors. Generally, one link is used to select the position of the
network. If the offset temperature coefficient is positive, the network
is connected to one side of the bridge; but if the error is negative,
the network is connected to the other side of the bridge. The first
thing the computer must determine is the sign of the error; it then
must open the appropriate link to connect the network. Both offset
temperature coefficient and offset can have either polarity and will
require a link network to set the right polarity before the value
is trimmed. Resistors are added to the perimeter (out of the strain-
sensitive area) to adjust offset and span by laser-trimming these
resistors after the chip has been tested. This, again, is taking
a function that is done by the user and making it a part of the
sensor chip, increasing the price/performance ratio still further.

Figure 12.5 shows the layout of a compensated and calibrated
pressure-sensor chip.


## 12.4  FUTURE DEVELOPMENTS

The future trends in silicon pressure-sensor development will be
determined by the consumer. If a number of different markets request certain features in the sensor, such as temperature-compensated
and calibrated devices, the development effort will go in that direction. As new markets emerge, other requests will be made, such
as multiple devices in one package or combined temperature, pressure,
and flow devices.

Traditional markets for silicon pressure sensors and microstructures are

- Automotive
- Medical
- Aerospace/military
- Process control
- Industrial control

There are several new, emerging markets for silicon sensors, including

- HVAC
- Consumer
- Computer peripherals
- Automation

Of these new markets, HVAC and consumer will probably have the biggest impact on the sensor industry.

*HVAC*:   HVAC (heating, ventilation, and air conditioning) applications require very low-pressure, low-cost sensors to monitor the airflow throughout an air-conditioning system. One system could use many sensors to get an accurate reading of the system performance. A low-pressure-differential sensor in the pressure range of 0 to 1.5 inches of water will have to be developed. The output probably will have to be a frequency or digital output, due to the distances between the computer and the sensor.

*Consumer*:   Microprocessors are being designed into almost all the new appliances, such as microwave ovens, washing machines, and dishwashers. As the trend continues, these devices will require sensors to get feedback from the analog world. Another advantage is that these products can be made more efficient, resulting in lower power consumption.

*Computer Peripherals*:   Another application is devices such as pressure sensitive joysticks and marking instruments (drawing on the CRT) that make wide lines when pressure is applied and narrow lines when little pressure is applied.

*Automation*:   A broad range of sensors is required to provide automation in the manufacture of goods. These include tactile sensors and pressure sensors to detect the stress applied to the workplace, as well as devices to detect whether, for example, the arm has picked up a chip, or to monitor the pickup arm of a pick-and-place machine.

## 12.4.1  Electronic On-Chip Integration

*Passive Compensation*

As previously discussed, the first level of integration would be to balance the sensor by adding a network on the chip to trim the offset to zero. Higher levels of integration might include temperature compensation of both offset and span. The final passive compensation would be to normalize the offset and sensitivity of the sensors to make them interchangeable.

### Active Compensation

The sensor requires two active circuits to operate: a stable reference and constant current or voltage source, and an amplifier circuit to amplify the 100-mV sensor output to a level that can interface other electronic circuits.

Adding the active circuits to the chip has both advantages and disadvantages. The more complex chip would require more engineering and tooling costs to get the device into production; but, on the plus side, the unit cost would be reduced in high volumes. This would work if one pressure range and one sensitivity were required. Low volume or multiple pressure ranges would be a problem.

Every level of integration on the chip will increase the number of masks and wafer-processing steps necessary to produce the sensors. The chip would cost more than an uncompensated chip, but the time and money spent on adding the compensation externally must be considered when deciding to add more functions to the chip.


### 12.4.2  Smart Sensors

Another technology trend is the incorporation of microprocessors into sensors to increase the power for performing different functions at the sensor level. The low cost of the digital components, combined with the necessity of temperature-compensating the sensors, created the first smart sensors. The developments were incorporated in process-control transmitters. The incorporation of the microprocessor allowed the performance of silicon sensors to be improved.


### 12.4.3  Other Sensors

Sensors other than pressure sensors are starting to use micro-machining technology. Accelerometers and other mechanical devices are being machined from silicon wafers. These sensors use the same basic Wheatstone bridge, which is strained as the acceleration is sensed by a mass machined from the silicon wafer. The mass is suspended by two thin silicon beams, and four resistors are implanted into the beams that sense the strain as the sensor is accelerated. Low-cost, high-volume accelerometers can be manufactured using the same basic techniques as used for pressure sensors.

# Bibliography

CHAPTER 1

1.  Harry N. Norton. (1982). *Sensor and Analyzer Handbook.*
    Prentice Hall, New York.


CHAPTER 2

1.  Harry N. Norton. (1982). *Sensor and Analyzer Handbook.*
    Prentice Hall, New York


CHAPTER 3

1.  Specifications and Tests for Strain Gage Pressure Transducers.
    (1982). *ISA Standard S37.3.* Instrument Society of America,
    Research Triangle Park, North Carolina.
2.  Electrical Transducer Nomenclature and Terminology. (1982).
    *ISA Standard S37.1.* Instrument Society of America, Research
    Triangle Park, North Carolina.


CHAPTER 4

1.  Harry N. Norton. (1982). *Sensor and Analyzer Handbook.*
    Prentice Hall, New York.

## CHAPTER 5


1. H. Reimann and Y. Fathi. (1986). *Simulation of Diaphragm-Type IC Pressure Sensors*. Solid-State Sensor Workshop, Hilton Head Island, South Carolina.
2. S. Timoshenko and S. Woinowsky-Krieger. (1959). *Theory of Plates and Shells*. McGraw-Hill, New York, pp. 347–348.
3. Hibbitt, Karlsson, and Sorensen, Inc. (1982). *Abaqus, A Finite-Element Code for Structural Analysis*, Providence, Rhode Island.
4. Research Triangle Institute, Technical Documentary Report No. ADS-TDR-63-316, Vol. 5, July 1964.
5. J. J. Wortman and R. A. Evans. (1965). Young's Modulus, Shear Modulus, and Poisson Ratio In Silicon and Germanium, *J. Appl. Phys. 36*:153–156.
6. K. W. Lee and K. D. Wise. (1982). Sensim: A Simulation Program for Solid State Pressure Sensors, *IEEE Transactions on Electron Devices, ED-29*:34–41.
7. S. Suzuki and Y. Yagi. (1982). Optimum Design of Silicon Pressure Sensor by Nonlinear Finite-Element Method, *Proceedings of Sensor Symposium* (Japan), pp. 163–165.
8. Y. Kanda. (1983). Optimum Design Considerations for Silicon Pressure Sensors Using a Four-Terminal Gauge, *Sensors and Actuators 4*:199–206.
9. K. Suziki, T. Ishihara, M. Hirita, and H. Tanigawa. (1985). Nonlinear Analyses on CMOS Integrated Silicon Pressure Sensor, *Proceeding of IEDM*, Paper 5.8.
9. Kurt E. Peterson. (1982). Silicon as a Mechanical Material, *Proceedings of the IEEE, 70, No. 5*:439–442.
10. K. W. Lee and B. E. Walker. (1986). Silicon Micromachining Technology for Automotive Applications, *SAE Sensors and Actuators*, Detroit, Michigan.
11. E. Svoboda. (1986). Passive Compensation of a Monolithic Silicon Pressure Transducer, *Proceedings Sensors Expo*, Chicago.
12. *Silicon Sensors and Microstructures*. (1988). Nova Sensor, Fremont, California.


## CHAPTER 6


1. Specifications and Tests for Strain Gage Pressure Transducers. (1982). *ISA Standard S37.3*. Instrument Society of America, Research Triangle Park, North Carolina.
2. Electrical Transducer Nomenclature and Terminology. (1982). *ISA Standard S37.1*. Instrument Society of America, Research Triangle Park, North Carolina.

## CHAPTER 7

1. R. Lee and D. Tandeske. (1986). Design Techniques for Low Pressure Sensing Applications, *Proceedings Sensors Expo*, Chicago.
2. *Pressure Sensor Handbook*. (1987). Sensym, Sunnyvale, California.

## CHAPTER 8

1. *Pressure Sensor Handbook*. (1987). Sensym, Sunnyvale, California.
2. *Silicon Sensors and Microstructures*. (1988). Nova Sensor, Fremont, California.

## CHAPTER 9

1. *Pressure Sensor Handbook*. (1987). Sensym, Sunnyvale, California.

## CHAPTER 10

1. *Pressure Sensor Handbook*. (1987). Sensym, Sunnyvale, California.
2. *Silicon Sensors and Microstructures*. (1988). Nova Sensor, Fremont, California.

## CHAPTER 11

1. *Silicon Sensors and Microstructures*. (1988). Nova Sensor, Fremont, California.

## CHAPTER 12

1. *Silicon Sensors and Microstructures*. (1988). Nova Sensor, Fremont, California.

# Appendix

BASIC RELATIONSHIP TO DETERMINE FLOW

$$Q = V \times A$$

where

Q = liquid flow through pipe.
V = average velocity of flow.
A = cross section area of pipe.

CONVERSION FACTORS FOR AREA

|  | Meter$^2$ | cm$^2$ | ft$^2$ | in$^2$ | circ. mil |
|---|---|---|---|---|---|
| 1 meter$^2$ | 1 | $10^4$ | 10.76 | 1550 | $1.974 \times 10^9$ |
| 1 centimeter$^2$ | $10^{-4}$ | 1 | $1.076 \times 10^{-3}$ | 1550 | $1.974 \times 10^5$ |
| 1 foot$^2$ | $9.29 \times 10^{-2}$ | 929.0 | 1 | 144 | $1.83 \times 10^8$ |
| 1 inch$^2$ | $6.452 \times 10^{-4}$ | 6.452 | $6.944 \times 10^{-3}$ | 1 | $1.273 \times 10^6$ |
| 1 circ. mil | $5.067 \times 10^{-10}$ | $5.067 \times 10^{-6}$ | $5.454 \times 10^{-9}$ | $7.854 \times 10^{-7}$ | 1 |

## CONVERSION FACTORS FOR UNITS OF VOLUME

|               | $cm^3$      | $inch^3$   | liter              | $foot^3$              | $meter^3$                |
| ------------- | ----------- | ---------- | ------------------ | --------------------- | ------------------------ |
| $centimeter^3$ | 1          | .061024    | $9.999\times10^{-4}$ | $3.5315\times10^{-5}$ | $1\times10^{-6}$         |
| $inch^3$      | 16.387      | 1          | .0163866           | $5,787\times10^{-4}$  | $1.6387\times10^{-5}$    |
| liter         | 1000.028    | 61.02545   | 1                  | .035316               | $1.000028\times10^{-3}$  |
| $foot^3$      | 28,316.85   | 1728       | 28.316             | 1                     | .028317                  |
| $meter^3$     | $1\times10^6$ | 61,023.74 | 999.972            | 35.3147               | 1                        |

## SPECIFIC GRAVITY OF COMMON LIQUIDS

| Liquid              | SG   | Liquid          | SG   |
| ------------------- | ---- | --------------- | ---- |
| Acetic Acid         | 1.06 | Mineral Oil     | 0.92 |
| Alcohol, Commercial | 0.83 | Muriatic Acid   | 1.20 |
| Alcohol, Pure       | 0.79 | Naphtha         | 0.76 |
| Ammonia             | 0.89 | Nitric Acid     | 1.50 |
| Benzine             | 0.69 | Petroleum Oil   | 0.82 |
| Carbolic Acid       | 0.96 | Phosphoric Acid | 1.78 |
| Carbon Disulfide    | 1.26 | Sulphuric Acid  | 1.84 |
| Fluoric Acid        | 1.50 | Turpentine Oil  | 0.87 |
| Gasoline            | 0.70 | Vinegar         | 1.08 |
| Kerosene            | 0.80 | Water Sea       | 1.03 |
| Linseed Oil         | 0.94 |                 |      |

## SPECIFIC GRAVITY OF COMMON GASES

| Gas           | SG   | Gas               | SG   |
| ------------- | ---- | ----------------- | ---- |
| Acetylene     | 0.92 | Hydrochloric Acid | 1.26 |
| Air           | 1.0  | Hydrofluoric Acid | 2.37 |
| Alcohol Vapor | 1.60 | Hydrogen          | 0.07 |
| Ammonia       | 0.59 | Nitric Oxide      | 1.04 |

(Specific Gravity of Common Gases, Continued)

| Gas | SG | Gas | SG |
|---|---|---|---|
| Carbon Dioxide | 1.52 | Nitrogen | 0.97 |
| Carbon Monoxide | 0.97 | Nitrous Oxide | 1.53 |
| Chlorine | 2.42 | Oxygen | 1.11 |
| Ether Vapor | 2.59 | Sulfer Dioxide | 2.25 |
| Ethylene | 0.97 | Water Vapor | 0.62 |

## TEMPERATURE CONVERSION FORMULAS

Kelvin = degree celsius + 273.15

Celsius = (degree Farenheit - 32) × 5/9

Farenheit = (degree celsius × 9/5) + 32

## PRESSURE VS. ALTITUDE

| Altitude (ft.) | Pressure (in hg.) | (psi) |
|---|---|---|
| -1,000 | 31.0185 | 15.235 |
| -500 | 30.9073 | 15.180 |
| 0 | 29.9213 | 14.696 |
| 500 | 29.3846 | 14.432 |
| 1,000 | 28.8557 | 14.173 |
| 1,500 | 28.3345 | 13.917 |
| 2,000 | 27.8210 | 12.665 |
| 3,000 | 26.8167 | 10.715 |
| 4,000 | 25.8418 | 12.692 |
| 6,000 | 23.9782 | 11.777 |
| 8,000 | 22.2250 | 10.916 |
| 10,000 | 20.5770 | 10.106 |
| 12,000 | 19.0294 | 9.346 |

(continued)

(Pressure vs. Altitude, Continued)

| Altitude (ft.) | Pressure (in hg.) | (psi) |
| --- | --- | --- |
| 14,000 | 17.5774 | 8.633 |
| 16,000 | 16.2164 | 7.965 |
| 18,000 | 14.9421 | 7.339 |
| 20,000 | 13.7501 | 6.753 |
| 22,000 | 12.6363 | 6.206 |
| 25,000 | 11.1035 | 5.454 |
| 30,000 | 8.8854 | 4.364 |
| 35,000 | 7.0406 | 3.458 |
| 40,000 | 5.5380 | 2.720 |

# Index